The noughties brought to book

Book reviews 2000-2009

Michael Gross

ISBN: 978-1-4452-7240-5

Table of contents

Preface:

Same decade, different books

Among all the reflections of the various cultural aspects of the noughties, I found a bestsellers list for the entire decade in the German news magazine Der Spiegel. As was to be expected, the fiction list is dominated by wizards, vampires and other fantasy fabrications, with only two or three "thoughtful" books making the top 20. Even the non-fiction list bears very little resemblance with the real world.

A very different list of books is assembled and reviewed in this collection. Most are about science – ranging from popular to academic – but there are also a few literary titles from ca. 2007 onwards, when I decided to review (almost) every book I finish on my blog (www.proseandpassion.com).

Most of the science titles I reviewed in Chemistry & Industry, either in the "long essay review" or in a "short essay review" format. In both cases I have had the opportunity to contribute my own thoughts on the subject matter, which is the reason why I think that at least some of these reviews may be worth collecting and rereading even when the books under review may no longer be in the shops.

I believe that the main reason why I chose to read and review these books in the last 10 years, rather than those that most readers seem to prefer, is that all of these books, including the novels, tell us something worth knowing about the world we live in. I should clarify that I have no problems with magical elements popping up in fiction and I am in fact a great fan of Latin American *realismo mágico*. However, I believe that these elements are only of interest if the work is constrained in some way by the real world.

So, just in case that after a decade of irrationalism there is anybody else left who has an interest (scientific or literary) in the world, I'm stitching up these musings into a book, to make sure they remain accessible to the presumably very small number of readers who may be interested in them. After more than a dozen books I have published commercially, this is my first experiment with self-publishing, so any comments and critiques are most welcome.

Oxford January 2010

michael gross

I. Long reviews

These "long essay reviews" were all written for the magazine Chemistry and Industry. I started out reviewing books about nanotechnology in an attempt to keep in touch with a fast-growing field. As the decade went by, the reviews spread out into life sciences and chemistry, and also covered a few biographies of scientists. As the name of the format suggests, I have on each occasion had the opportunity to discuss my own thoughts on the subject matter as well as the content of the book under review.

Technology of our time

Nanotechnology
Michael Köhler, Wolfgang Fritzsche
Wiley-VCH

Originally published in Chemistry & Industry, No 15 (Aug. 2004) 22.

During the last ten years, the concept of nanotechnology has swiftly moved from the realm of futurology into the real world. Palpable progress in application-oriented fields such as surface patterning, paper-thin display technology, molecular computers, etc. is reported every week, and the manufacturers of computer chips now create nanostructures in millions of copies. While the rash of "nano"-labelled research institutes and projects could be dismissed as a fad, the results clearly indicate that nanotechnology has arrived. It is happening here and now.

Accordingly, the nature of the books on this topic is evolving. In the beginning there were the "vision" books by Eric Drexler and his school of thinking, where speculation on what might be possible was dominant. Then came a blend of real life science and crystal-gazing, including my own effort, Travels to the Nanoworld (attributed to "M. Cross" by Köhler and Fritzsche). Now that nanotechnology is part of our everyday science and technology, we really need even more practical books, such as student text books and "How to" guides for the growing number of non-specialists who will be exposed to nanotech issues in their working life, and for the wider public.

The present book from physical chemists Michael Köhler and Wolfgang Fritzsche sails under a fairly general title, but the prohibitive price tag clarifies that it is not addressed at a popular audience. Its style, both in text and illustration, is that of a text book. The authors take great care to provide a comprehensive coverage of all underlying general principles before descending into anything as lowly as a real-world example.

Thus, the first three chapters are solely painting the background for the picture that we are hoping to see later on. Chapter 4, finally, promises to come to the meat of it, as it carries the headline “Preparation of Nanostructures”. But alas, under 4.1, it’s “principles” again, and under 4.1.3 there are even more principles. In 4.2, finally, we get a glimpse of real world structuring techniques which the reader may have seen on the pages of Science or Nature, such as Dip-Pen Nanolithography (DPN). The third and fourth section of the chapter then roll out the whole arsenal of nanostructuring techniques, devoting one to two pages each to methods like electron-beam lithography, direct writing, extreme UV and X-ray lithography, and so on.

Now these sections should convey the exciting news that we are now able to manipulate nanometre structures with a range of different techniques, and the resolution is getting better all the time. And they should be giving students an idea of how it is actually done. Sadly, neither the excitement nor the practical side of things come across very well. Part of this may be due to the overly academic style which makes even this reasonably educated reader feel like Winnie the Pooh listening to Owl. For instance, to “explain” the use of two-directional deflection of electron beams, the authors helpfully specify that it “is known from cathode tube display technology.” Now there are two types of readers. One who understands this explanation, but doesn’t have a need for it. The other will not understand a word of it but would have benefited immensely if the authors had said something like: “just like in an ordinary TV screen.”

The fifth chapter, cryptically titled “Nanotechnical structures,” deals with molecular, i.e. bottom-up approaches to creating nanostructures, including DNA, dendrimers, supramolecular chemistry, etc. It introduces some three-dimensionality into the book which otherwise deals mainly with surfaces. The next chapter, however, falls back to the surfaces and to methods used to characterise them. Chapter 7 presents the action side of nanotech, including actuators, sensors, transducers, transistors and so on. The final chapter very briefly covers general aspects of the design of nanotech-based complex systems.

Much as I admire the authors’ thoroughness and their attempts to outline general principles in a field that admittedly appears less strictly organised in my own writings,

I am slightly worried about the potential readership. Assuming that this book is to serve as a text book for chemistry or physics undergraduates or graduates who consider specialising in nanotechnology, it will be for the consumption of a generation whose attention span is famously dictated by the length of video clips or even TV ads. Seeing that I had trouble cutting my way through the hedge of generalisations and principles that surrounded the sleeping beauty of really exciting nanotech research, I am worried that too few students will be willing to study this book from cover to cover.

Reassuringly though, there are around 100 quite instructive diagrams, so any student who finds the text too challenging can still benefit from flicking through those. There remains a need for a generally accessible "How To" guide to nanotechnology, both for specialists and for the non-specialists who will increasingly need to know about the field that is becoming the defining technology of our time.

A big book of small technology

Springer handbook of nanotechnology
Bharat Bhushan (ed.)
Springer Verlag, Berlin, 2004

Originally published in Chemistry & Industry, No 6 (21.3. 2005) 24-25.

Nanotechnology has finally arrived in the real world. Devices using sub-micrometer structures for digital data processing, e.g. data projectors, are already in the shops. In the research labs, biotechnology and microelectronics merge into one, as researchers successfully connect neurons to electrical circuits, produce patterns of living cells using commercial inkjet printers, or use DNA arrays that can be addressed at ever higher resolution. Clever DNA constructs can walk along a track, grasp and release molecules, or diagnose disease. Gene sensors that are sensitive to single point mutations are already available, and further diagnostic tools and novel therapeutics based on these new technologies are on the horizon.

A lot more is yet to come, once the emerging technologies such as carbon nanotube electronics have outgrown their teething problems. We have grown so used to the perpetual computer revolution with its rate of progress undiminished over the last three decades that we may be taking rapid change for granted, but it is obvious that nanotechnology has already started to take over the role of the key technology for the continuation and diversification of that progress, which will continue to change our lives over the years to come.

Another indicator of the nanotech revolution is the flood of written words that came with it. One might dismiss the fact that the heir to the British throne is now voicing his concern about nanotechnology and the grey goo scenario (in response to which Eric Drexler voiced his regret about popularising that scare story) as a quirk, but it also indicates that, for better or worse, nanotechnology has finally reached the minds of non-scientists. At this point it is all the more important for the scientific community to display an objective view of what is possible and desirable, and what is not.

Scientists seem to have responded to the communications challenge by producing scores of new books. The year 2004 saw many new titles containing the prefix "nano-", but did the books do the nanotechnology revolution justice? Have they succeeded in giving an accurate overview of the most important technological change of our time? Many of the new books remain limited to a specific subdiscipline, which authors and editors try to sex up by using various nano-words. After reviewing a few of them, I concluded in this very magazine (2 August 2004): "There remains a need for a generally accessible how-to guide to nanotechnology both for specialists and non-specialists ...".

Now, with the Springer Handbook of Nanotechnology, there is a bulky volume with the clear ambition to fulfil this need. With over 1200 pages, 972 illustrations, 90 authors, and 38 chapters, and an accompanying CD containing the full searchable text, it looks like a book matching the breadth and importance of its subject matter. So is this the book we have been waiting for, the one that does nanotechnology justice?

Let us first look at the large scale structure before zooming in to the details. On the largest scale, the book can be seen as a triptych of broad canvasses. The chapters covering nanotribology, nanomechanics, and thin films (the issues most closely related to the editor's own research interests) take the central place, although with 13 chapters and approx. 400 pages they do not consume an excessive amount of space. They are framed by a general and detailed introduction into nanotechnology (16/500) and a section on applications and implications (9/250). While the editors have subdivided each of these three parts into two sections, resulting in a sequence of six sections labeled A-F, my impression is that the tripartite structure would be more user-friendly. One could imagine the book, which is almost too heavy to lift with untrained arms, coming out as a portable and affordable student edition of three paperbacks in a nice cardboard box a couple of years down the line.

The chapters, especially those in part I (my division), are surprisingly readable and well-illustrated. A lot of editorial effort must have gone into streamlining the text produced by 90 different authors to the extent that one could almost forget that this is a multi-author monograph. Moreover, the graphics for all chapters have obviously been redrawn to create a unified and quite pleasant visual style. Each chapter is fronted by a page containing the title, abstract, and detailed table of contents. Every chapter has a full list of references, typically around 100 of them, but chapter 3 has 352. Obviously, as such a volume takes years of preparation, the references are not quite up to date, but the years up to and including 2003 appear to be covered reasonably well.

The streamlining and consistency work done so well by the editors may have gone over the top with respect to the listing of the authors, their affiliations, and acknowledgements. In fact, no author-related information is found anywhere near the chapters. While creating the illusion of a volume that was produced in one go by an omniscient creator (why not call it the nanotech bible?), this editorial policy brings some inconvenience to the reader as well.

Suppose you are interested in carbon nanotubes and want to get in touch with the authors of the excellent introduction to nanotubes presented in chapter 3. Who are the authors? The chapter won't tell you. The full table of contents (page XXI) doesn't tell you either, nor does the full and detailed table of contents on page 1171. Handy hint: For a full listing of the authors of chapter 3, you have to consult the partial table of contents offered at the beginning of the section A, of which the chapter is a part, on page 7. There you will find the names of the 8 authors of the chapter, all located at Toulouse, France. But neither address nor biographical information. To find their addresses, you could turn to the list of authors at the front of the book (page XIII), where all 90 authors are listed alphabetically with their full addresses. But which one of the eight should you try to contact? To find out who's the professor and who's the graduate student, you need to remember all those eight names and look them up again in the second alphabetical list of authors at the end of the book, on page 1155. This list has the addresses again, plus a photo and short biography of each, so you can pick whoever is most senior or looks most approachable. Oh, and if your are already a collaborator of the group and want to check whether they have acknowledged your contribution, you need to turn to page 1153. Call me old-fashioned, but I think this information (except for the photo and bio) belongs with the chapter.

But obviously these editorial quirks, along with some linguistic ones and the embarrassing deference to US institutions including “former president William J. Clinton” (not to be confused with the Bill C. we all know) are insignificant glitches measured against the ambitious scope and sheer bulk of this volume which remains the best and most comprehensive reference on nanotechnology that I have seen so far.

Perfectly formed

Nanochemistry. A chemical approach to nanomaterials
Geoffrey A. Ozin and André C. Arsenault
Royal Society of Chemistry 2005

Originally published in Chemistry & Industry, No 15 (7.8. 2006) 25-26.

Nanotechnology is about controlling matter at the molecular scale. Now isn't that the definition of chemistry, too? The subtle difference is that in chemistry, we often exert our experimental control in a very non-specific way, e.g. by heating or mixing samples that might contain 10^{20} molecules or more. At some time, some of the molecules will do what we want them to do, and others will follow suit. Which molecule will react at what place and what time isn't particularly important to the chemist, who sees molecules as faceless members of an unimaginable multitude.

The nanotechnologist, in contrast, might be described as a chemist who has given up his or her liberal laissez-faire approach and turned into a control freak. The chemist who wants individual molecules or atoms to jump through a specific hoop at a specific time, who wants them to form a specific arrangement in a pre-defined location, to create meaningful patterns (such as, err, the letters IBM or a smiley face) or functional elements such as molecular transistors or motors. Nanotechnology is the adventure of chemistry that doesn't seek safety in numbers.

What, then, is "nanochemistry" supposed to mean? If nanotechnology and chemistry are so closely related already, is there space for an interdiscipline between them? Would it make sense to reward the control freaks among us with a new chemical discipline to be lined up alongside the traditional subjects of organic, inorganic, physical, and biological chemistry?

As the subtitle of this book helpfully clarifies, nanochemistry is named the wrong way round. It is not so much trying to define a new kind of chemistry, but creating a specific, namely a chemical approach to nanotechnology. Chemical nanotechnology makes a lot more sense than nanochemistry, as there are indeed several different ways of approaching the challenges of the nanoworld. There are a few access roads from physics, some from biology, and there is the self-assembly, bottom-up approach using the methological repertoire of a chemist (including, for example, supramolecular chemistry, organic synthesis, and the physical chemistry of colloids and thin films).

So here we have the work of two chemists who describe their specific approach to nanomaterials in a textbook format. Based on a course he developed at the University of Toronto, Geoffrey Ozin wrote this text with his student, André Arsenault. The resulting tome has thirteen chapters and seven appendices, and it comes lavishly equipped with many full-colour illustrations, some 2,000 references, lists of questions "designed to inspire creative and holistic thought," and home-made cartoons.

The chapters essentially appear to stand side by side, and -- apart from the first and the last two -- they can probably be consumed in any order or selection. As in most nanotech books, the nanotubes and -wires take the largest chunk of the space. "Biomaterials and Bioinspiration" turn up relatively late in chapter 10, so Nature doesn't get that much credit for having come up with nanoscale machinery some 3.5 billion years before we did. The issue of the possible cytotoxicity of nanoparticles (which may be quite distinct of the toxicity of the corresponding bulk materials) has only begun to be explored in recent years, after the authors completed the main text. Therefore, it appears in one of the appendices.

Much like the arrangement of the chapters doesn't seem to be supporting any particular storyline, the numbered sections -- sometimes imaginatively titled, but often too short -- within each chapter aren't connected very well either. Even within the sections, the sentences stating separate facts seem to follow each other in a *staccato* rhythm, rather than linking up to a *legato* tune.

All in all, this book certainly has all the interesting bits and pieces covered, and it can be reassuringly authoritative in its presentation of facts, which I am sure are absolutely correct, but it turns out to be surprisingly difficult to read more than two or three pages of it in one go. I suspect that the lack of storytelling, of joined-up presentation, of fitting the pieces together to a big picture, makes this book more difficult to read than it needed to be based on the complexity of the subject matter alone. Maybe the authors are expecting that students will effortlessly join the dots for themselves.

This kind of presentation found in many nanotechnology books is more than just a cosmetic problem because nanotechnology is by its very nature interdisciplinary, as it involves methods borrowed from physics, chemistry and biology, and has ambitions that reach deep into medicine and engineering, to name but a few of the traditional disciplines it spans. With this breadth of the topic comes a communications challenge, as readers with a background in any of these disciplines will want and need to understand what experts from the other parts of the nano spectrum have to say.

Following several attempts by physicists and application-oriented people, some of which I found to be difficult or even unreadable for anybody from outside the authors' specific sub-discipline, this textbook of chemical nanotechnology demonstrates that chemists face the same challenge as the physicists and biologists that have gone before.

I would have thought, however, that it shouldn't be that difficult for chemists to reveal some kind of logical and easily accessible structure in the field of nanotechnology. I've always favoured the "nature vs. technology" juxtaposition, but a "bottom-up" approach rising from atoms via small molecules to macromolecules and macromolecular assemblies would probably work just as well and would appropriately mirror the "bottom-up" synthetic strategy which is characteristic of the chemical approach to nanotechnology, as opposed to the top-down approach represented in further miniaturisation of micromanufacturing techniques.

There must be some chemists out there who have got the facts at their fingertips (I lost the overview some time ago and turned to reviewing nanotech books in an attempt to catch up!) and who can link them up into a compelling story. And please don't call it "nanochemistry." That word makes me shiver.

Heady metals

Neurodegenerative diseases and metal ions (Metal ions in life sciences, vol. 1)
A. Sigel, H. Sigel, and R. K. O. Sigel, eds.
Wiley, 2006

Originally published in Chemistry & Industry, No 5 (12.3. 2007) 30.

Ever since the Romans allegedly poisoned themselves with the lead contained in their drinks, the toxicity of metals and especially their effects on the brain have been a major concern. For example, I remember a time in the 1990s, when the discovery of aluminum in the amyloid plaques from the brains of Alzheimer patients triggered fears that traces of this metal rubbed off from cooking pots might cause the disease.

Public perception of such threats is often operating on the simplistic level assuming that metal is ingested, goes to the brain, then makes you go soft in the head. In reality, however, toxins that we ingest don't normally cross the blood-brain barrier or enter nerve cells. If they do, there may be a biological reason for it. And if a metal ion has a biological role, there is likely to be a complex system in place that allows and regulates its transport and keeps its concentration in the cell constant (homeostasis).

To put it as simply and bluntly as possible, if you have too much copper in your brain cells, it's probably not because you have eaten too much copper, but more likely because there is something wrong with the complex copper-regulation network. Therefore, the links between metal ions and brain disease are diverse, complicated, and far too little understood.

With this volume, the first in a new series on "Metal ions in life sciences," the three Sigels (one assumes they are members of one family) present a multi-author monograph covering this issue in great detail. The main topics covered include: copper and its relation to amyloid diseases; iron / Huntington's and Parkinson's disease; zinc; aluminum; and the neurotoxicity of Cd, Pb, Hg.

To pick just one of these areas as an example, the interaction of copper with prion diseases has been a hot research topic over the last few years. Copper ions are vital but also potentially dangerous to the cell because of the redox chemistry they can engage in. Accordingly, a whole range of proteins have been identified that take part in copper homeostasis in one way or the other, including transporters and "copper chaperones."

There has been a lot of circumstantial and structural evidence suggesting that the "healthy" version of the prion protein is involved in copper traficking and homeostasis in some way, as reviewed in chapters 3 and 4 of this book. Earlier this year, researchers at the Free University of Berlin demonstrated a two-way communication between the metal and the potentially infectious protein. The presence of correctly folded prion protein is important for the mainenance of copper homeostasis. If the protein is removed from the cell, a change in external copper concentrations will result in an unhealthily large change in internal concentrations as well.

Conversely, if the homeostasis of copper and manganese ions runs out of control, certain conditions may also trigger the pathological misfolding of the prion protein into its infectious form. Thus, it appears to be crucial to have a healthy balance of protein and metal concentrations to keep the cell on the straight and narrow path to well-being.

This book is impressive as a collection of very thorough and comprehensive reviews on a very timely and important, possibly even still underappreciated subject area. Medical experts should consult it to brush up on the chemistry behind the pathologies they are confronted with, and bioinorganic chemists will appreciate that it highlights the importance of their field for real-life problems.

However, I have recently been confronted with a researcher in a related field who quite fiercely argued that in the times of fast electronic journals, the slow and labour-intensive publications of such reviews in book format might be entirely pointless. While I don't quite subscribe to this view, the "added value" of the book format needs to be looked at.

In this particular case, the exceptionally tight cohesion and high importance of the subject matter, seven metals that are involved in major neurological problems, seem to argue in favour of a book publication, and the present volume does this subject matter justice. On the other hand, one hopes that rapid progress is going to be made in many of the areas covered by the book, so its useful shelf life may turn out to be rather short.

If one goes to the trouble of bundling reviews in a book, even though they might have been more easily and rapidly published in electronic journals, one should consider ways of including some bonus features that provide added value. The subject index is an obvious one, and with 27 pages of detailed indexing, this book doesn't disappoint. An introduction of more than 5 pages might have been useful, and it might have included the general background knowledge of what biological functions these metals generally have (a summary of R.J.P. Williams's brilliant books on that matter might be useful), how they get to the brain, how homeostasis works, and so on. It should set the scene for the encounter between metals and the components of the nerve cells in a more general, lay friendly way, allowing people with specialist knowledge in either chemistry or neurology to start from the same level.

Finally, as the seven metals mentioned here have accompanied human history for a while (with the exception of Al, which is difficult to produce), a chapter on the history of their use, misuse, and toxicity might have been interesting. For instance, I wanted to know whether the Romans really did become stupid from consuming too much lead, or whether that is an urban myth, but there is no reference to Romans in the index. Maybe there is a popular science book waiting to be written here, to plumb these historical depths and contemporary challenges (apologies for the leaden pun!).

Oh, and for those who would like to collect the series (the Sigels' previous endeavour ran to 44 volumes!), I can reveal that the second volume will be about "Nickel and its surprising impact in nature," and the third one about cytochrome P450 proteins.

Being Sage

JD Bernal: The sage of science
Andrew Brown
Oxford University Press, 2005

Originally published in Chemistry & Industry, No 1 (15.1. 2007) 28-29.

Since his student days at Cambridge, John Desmond Bernal had the reputation of knowing everything, hence the nickname "Sage" that was to accompany him for the rest of his life. His key role in the science of the 20th century was that he layed the foundations for the X-ray crystallography of biomolecules, a method which by now has elucidated the three-dimensional structures of around 40,000 proteins at atomic resolution. In this field, he inspired research projects that were to be rewarded with numerous Nobel prizes, handed to researchers including Dorothy Hodgkin, Max Perutz, John Kendrew, James Watson, Francis Crick, Aaron Klug. One gets the impression that the Nobel committees were so busy handing out the gongs to his disciples and followers that they somehow forgot to honour the inspirational source of their work.

Sage's wisdom and knowledge stretched far beyond this chosen field, including not only science and technology, but even remote corners of the history of art, politics and history. He was aware of his limitations, however. Asked whether he knew everything, he reportedly thought for a while, then replied: "I know nothing at all about the fourth century in Romania."

On top of his role as a 20th century Leonardo, Sage also was a tireless political activist on the political stage, fighting for the causes of communism and the peace movement after WW II. And then he still had some time and energy left to juggle a rather complicated love life, with several women regarding themselves as his "wives" at any given time.

Thus it appears that there is material for three biographies here, reporting the lives and times of the Wise Man, the Activist, and the Don Juan. Weaving these into one story that still makes some kind of sense and fits into a book format is an enormous challenge. Not including the 68-pages essay by Dorothy Hodgkin for the *Biographical Memoirs of Fellows of the Royal Society* published in 1980, this is (to my knowledge) only the second attempt to turn Bernal's life into a book. The first such endeavour, by Maurice Goldsmith, ended up being boycotted by several of the crucial witnesses of Bernal's life. As Goldsmith was denied access to Bernal's papers, his book, also published in 1980, had to remain an incomplete effort.

Now, a quarter of a century later, Andrew Brown has made a new attempt to squeeze the chaos that was Bernal's life between a pair of book covers. He has enjoyed access to the archive of Bernal's papers, except for one batch of love letters that has to remain sealed until 2021, but which are considered unlikely to contain any Earth-shattering revelations.

Brown tries to bring some order into Bernal's life by using a chapter structure that has both thematic and chronological elements. If there is any crystallographic symmetry in the book, it is provided by the central four chapters on Bernal's role in the second World War, which are flanked by eight chapters on his early years and an equal number on his later years.

One of the key merits of Brown's biography is that he seems to have succeeded in sorting out what exactly Bernal's role was in run-up to D-day. There had been conflicting evidence from Bernal's diaries and statements of some of the contemporaries, which had remained unresolved in previous efforts. For those who want to dig into this thorny issue more deeply, Brown offers a post-script with details of how and why he came to the conclusions stated in the text.

For the historically less ambitious reader, however, the question is whether Brown succeeds in bringing to life one of the most influential scientists of the 20th century, and whether we can maybe even learn something about his science by reading about his life.

On these two counts I found myself feeling just ever so slightly disappointed. Not to be misunderstood, I should point out that I stand in awe and admiration before the sheer courage, effort and determination needed to tackle the biography of such a larger than life figure. Thus, any criticism is not meant to diminish the biographer's merit. Rather, I would consider it possible that writing a perfect biography of such a complex human being is incompatible with the laws of physics.

Having said that, I must vent a couple of moans here. As I had already been familiar with many members of the cast from Georgina Ferry's brilliant biography of Dorothy Hodgkin, I didn't feel that the 562 pages bulk of this biography provided me with an equivalent amount of enlightenment on Bernal and the numerous contemporaries mentioned. Bernal himself, obviously, comes to life through the large number of anecdotes told about him, yielding the picture of a somewhat distracted genius, who needs others to organise his life for him, but is extremely generous in sharing his inspirational ideas with all and sundry.

And yet, the biographer often appears to be unable to understand what really made the great man tick. How did he choose the topics he developed an interest in, how did he choose his wives (seeing he can't have had much time to go looking for them), how did he maintain friendships despite a frantic lifestyle, why did he stick with hardline communism? As many such questions remain unanswered, it appears that even a biographer who has read all of Bernal's paperwork cannot necessarily read his mind.

Secondly, most of the supporting cast members drift in and out of the story without allowing us too much insight into their minds either. There are in fact so many of them that a short dictionary of names in the appendix would have been useful, especially as some of them change their names half way through the story (e.g. Professor Lindemann all of a sudden becomes Lord Cherwell). Celebrities like Picasso and Pablo Neruda get name-checked, but we are never told how close their ties with Bernal were, and whether these bridges across the "two cultures" divide served any cross-fertilisation beyond the famous Picasso mural drawn on a Birkbeck wall. Intriguingly, Soviet leader Nikita Khrushchev is one of the very few characters of the story whose image in my mind changed substantially from reading this book. I found it reassuring to learn that Khrushchev had long and deep discussions with someone like Bernal and that he admitted to being scared by the Cuban missile crisis.

Maybe the underlying problem is that, in order to understand Bernal, his science, and his friendships in depth, one would have to possess a mind at least as great as Sage himself. So I may be asking the impossible when I clamour for a guided tour of his inner world. Mere mortals like us will have to remain clueless as to what it was like, being Sage. In the meantime, Brown's book is a good approximation to an understanding of the life of Sage.

Musical metaphor

The music of life: Biology beyond the genome

Denis Noble

Oxford University Press

Originally published in Chemistry & Industry, No 9 (7.5. 2007) 29-30.

The largest jigsaw puzzle available from my local toyshop has 18,000 pieces. Imagine you were mad enough to try an even bigger one. Imagine 25,000 pieces tipped out on the table. No matter how good your jigsaw-solving technique, it is fairly safe to assume that this task will take you an enormous length of time to complete, and that most of that time will be spent not on putting pieces together but on picking up individual pieces, looking at them, sorting them into categories ... Only after a substantial period of looking and sorting will you be able to begin fitting some pieces together, probably those belonging to some well-defined subset, such as the borders of the image or some conspicuous feature.

Similarly, biochemists have spent most of the 20th century investigating and sorting individual pieces of the puzzle that is the living cell, with its many thousands of genes, proteins, carbohydrates, and smaller molecules. Only recently, around the turn of the millenium, have researchers entered the phase where they can hope to put the pieces together to get the big picture, using computer simulations to fit the wealth of information about the individual components, generated by a century of busy reductionist work, into a meaningful whole.

Since then, these efforts have been focussed in the new subdiscipline called systems biology. Noble's "little book" -- as he describes it himself -- is something like a manifesto for this new branch of science, trying to persuade both the experts and the intelligent lay reader that the systems perspective requires a whole new way of thinking about biology.

As somebody who has spent most of his life looking at individual pieces of the big jigsaw, Noble is far from rejecting reductionism. However, what does seem to bother him a lot in traditional reductionist thinking is the assumption that the chain of command must always start at the smallest unit, so the genes drive the proteins that drive the cell that drives the organism.

Systems biology, according to Noble, is not just about putting the pieces together to represent the whole system, it's also about realising that functions can arise on the system level, on the molecular level, or on any other level in between. While it is true that genes determine the amino acid sequences of proteins, it is just as true that certain proteins, including the transcription factors, decide whether a given gene gets to determine anything at all.

Therefore, he advises caution for the use and misuse of Dawkins' "selfish gene" metaphor, which has left a wide audience with the impression of the gene being in control of everything. Instead, he chooses the lead metaphor of orchestral music, where many different participants need to play together. (Although the cover of the book shows only one instrument: to the delight of the young cellist in my family it is a hybrid between a cello and a butterfly.) The genes in his view are no more than a CD that carries a recording of that music, and that, given the right cellular environment, can reproduce the whole complex network of the musical piece. Genes without an organism able to express them, are not really alive, just as a CD cannot produce music without a CD player.

Noble's musical metaphor reaches its limits when he has to admit that there is no central conductor, nor indeed an opera theatre in the brain, into which the perceptions of the "self" are projected. He seeks refuge in a wide variety of other imagery, from omelettes to Chinese calligraphy.

As I understand it, however, his answer to the old conundrum "What is Life?" is most closely related to the view of the network people who see life and even consciousness as an "emergent" property of networks that have reached a threshold degree of complexity. Long before this kind of thinking became fashionable, the chemist and writer Primo Levi pre-empted this view with a short story describing how a telephone network (the largest and most complex communications network before the internet came along) passes this threshold when two regional companies combine their networks. In this story, the larger network suddenly starts to form opinions on who should talk to whom and ends up interfering with the lives of its users.

The music of life is an important book on an important and very recent paradigm change in the life sciences. In spite of its small and slim appearance, it contains deep thoughts throughout, which is why it cannot be read as quickly as the small format might suggests. For every minute of reading the text, you may need a few more minutes to think about it and possibly to readjust your views on science, philosphy, and everything in between. If you take your time and approach the new systems view with an open mind, this book is a good introduction into this new way of thinking and a rewarding experience throughout.

One of the many lessons to be learned from this book is that all metaphors used to describe life can only be trusted to a very limited extent. That includes the giant jigsaw I used above. While a living cell doesn't necessarily have a larger number of different pieces than a large jigsaw, it is made immensely more complex by the possibility that any single piece might influence any other, or indeed entire groups of pieces, and vice versa. This is why life is much more interesting than any jigsaw puzzle and indeed than any other metaphor used to describe it.

Fish tales

When a gene makes you smell like a fish ... and other tales about the genes in your body
Lisa Seachrist Chiu
Oxford University Press 2006

Originally published in Chemistry & Industry, No 10 (28.5. 2007) 31-32.

With the complete human genome safely enshrined in searchable databases, the science of genetics has progressed to an entirely new level of analysis. Twenty years ago, when I spiced up my chemistry studies by attending a course on general and molecular genetics, the study of mutations (occurring naturally or produced artificially in laboratory animals) made up everything that could be known about our genes. An amazing body of knowledge had already been accumulated on simple human diseases caused by single mutations, such as sickle cell anaemia or haemophilia.

But nothing whatsoever was known about genes that might predispose people to autism, heart disease or mathematical ability. In other words, geneticists knew everything about the few monocausal genetic disorders and nothing about the many multicausal ones, which on the whole affect a larger number of people. In a rather unfortunate way, this concentration of a few atypical but easy targets led to misconceptions in the public eye. Having been told that a simple amino acid exchange led to sickle cell anaemia, and that an extra copy of one chromosome produced Down's syndrome, people came to expect simple explanations for more complex traits of individual human beings. Why hasn't the gene for homosexuality or brilliance at chess been found yet? -- many may have wondered.

Now, in 2007, we are a little bit wiser. Geneticists may have used up to the last drop the reservoir of easy targets, but the good news is that with the genome sequence, the increasing understanding of epigenetics (i.e. how the cellular environment influences gene expression), and the toolkit to match genetic diversity to the wide range of medical phenomena observed, scientists are now in a position from where they can start to tackle the really tricky issues of human genetics.

Now where is the gene that makes you smell like a fish to be placed in this landscape? As the catchy title suggests, Lisa Seachrist Chiu's book is rooted in the old genetics of the "low-hanging fruit," as the author puts it, i.e. in the study of disorders caused by mutations of a single gene. However, she takes this traditional starting point of mutations and their strange effects to gently lead the reader towards the things that matter so much more today, including genomics, haplotype and copy number diversity, epigenetics, and so on.

Some of her "tales about the genes in your body" are well-hung and much-repeated tales indeed. I, for instance, had been familiar with the stories of phenylketonuria and Huntington's chorea, and I had heard of many others, such as the werewolf gene and the taster phenomenon. Fortunately, the author always goes beyond the 1980s level and connects the mutant scare stories to 21st century genetics, reporting recent results from genomic and post-genomic studies along the way. The result is a sequence of -- often dramatic -- stories that are fascinating both for their compassionately told connection to human fates (amongst which the fishy smell is one of the milder afflictions), and for their wealth of intriguing scientific detail.

The fishy smell story of the title -- which was new to me -- is about a single gene metabolic disorder. In the disease officially known as trimethylaminuria (TMAU), the body fails to break down trimethylamine, a byproduct of protein metabolism. This smelly substance is then secreted in urine, breath and sweat, producing a rather strong, unpleasant odour. Although scientists have indentified the gene behind the problem, the causation is not quite straightforward, as the smell may only start to emerge well after puberty, and the trigger that determines the onset time remains unknown. TMAU often remains undiagnosed. It may affect as many as one percent of the world population, and as yet there is no cure.

Along with the compelling stories, the book offers a unique set of watercolour illustrations -- sadly printed in black and white only -- contributed by the author's mother. The comprehensive bibliography covers research up to 2005. And there is a "genetics primer" of seven pages as a crash course for those struggling with the elementary concepts of this discipline.

Which brings us to the readership. Will those people who need to refer to the genetics primer pick up the courage to read the book at all? I really hope they do, because there are few popular science books that combine the strengths of human interest story telling and the nerdy but necessary obsession with scientific detail quite as nicely as this one does. Furthermore, a book like this might have very easily ended up as a freak show of 20th century genetics, so it should be applauded for building a bridge between the old science of genes that go wrong and the new science of our genome.

In a perfect world, intelligent lay readers would use this book to graduate from genetics and enter genomics, so they could then move on to develop an interest in the postgenomics, epigenetics, proteomics, and system biology (see the previous chapter, "Musical metaphor") and other exciting things going on in this new century of science, which eventually may also be able to help those people (including the fishy smell patients), for whom the identification of "the gene" responsible for their condition hasn't resulted in a cure, because even simple genetic diseases may depend on more things than geneticists have dreamed up in their philosophy.

In the real world, however, I am wondering how many readers will endeavour to embark on this voyage of discovery.

Molecules in motion

Middle world
Mark Haw
Macmillan / palgrave 2006

Originally published in Chemistry & Industry, No 11 (11.6. 2007) 31-32.

In a perfect world the laws of physics would tell us everything we ever needed to know to understand all things large and small, from quarks to galaxies. As many of the inhabitants of our real world will know, it doesn't quite work out like that. There is one relatively straightforward set of physical laws for the Newtonian world of planets and apples, and a different, more challenging set for the quantum world, governing the behaviour of atoms and their constituent parts.

What is less well known is that in between these two there is a third domain, where neither Newton's nor Heisenberg's laws are very useful. It is the world of the molecules of life, which typically measure a few nanometres, which is why I usually refer to it as the nanoworld. Our author prefers to call it the middle world, so I will use his terminology here.

The main reason why the middle world is different from its two neighbouring worlds, Haw argues, is Brownian motion. Essentially, there is nothing special about Brownian motion, it's just the random back and forth of molecules that comes with thermal energy at all temperatures except absolute zero. Molecular motion doesn't trouble the movements of planets and footballs all that much, because it happens on such a small scale, and if you do the integration of all molecular motions around a macroscopic object, they will average out to zero.

However, if you peer down a microscope at very small objects, as Scottish botanist Robert Brown did back in 1827, you will see that thermal motion creates havoc in the middle world. Objects like the pollens that Brown studied are kicked about randomly and remorselessly by the molecules impacting on them.

Considering this, it is quite surprising that the nanoscale machines of life can work at all. Try riding a bicycle when you're bombarded with a hail of tennis balls (and maybe the odd basketball as well) from all sides, it's bound to be tricky. What science has found out quite recently is that the machinery of life doesn't just tolerate the random molecular bombardment that is the hallmark of the middle world. Instead, it very cleverly uses Brownian motion to drive the essential movement processes of life by biasing the randomness into the preferred direction. The energy that was previously thought to fuel cellular motors (in a naive scaled-down analogy to Newtonian motors) is now considered more likely to be invested in the biasing process. (The second law of thermodynamics quite firmly insists that we cannot get motion out of thermal randomness, we need to invest energy into selecting the "right moves.")

Mark Haw, a materials scientist at the University of Nottingham who has spent many years studying Brownian motion, has delivered a flamboyant plea in favour of a wider appreciation of the middle world. He starts with Robert Brown's observation of the pollen dancing in a drop of water, then steps back in time to the ancient Greek origins of the atom (the excuse being that Brownian motion offered one of the first experimental proofs of their existence), and then follows a vaguely chronological story line through the history of science, visiting Newton, the origins of thermodynamics, Boltzmann and Einstein, among many others. After a wild ride touching nearly every subject under the Sun, he manages to finish off at the very beginning, namely the origin of life.

It is extremely rare for a working scientist to let his hair down and go wild, combining a scientific subject with literary and philosophical ambition, so we should congratulate Haw on his sheer courage and determination to pull this off. However, my enthusiasm for his historical rollercoaster faded gradually towards the end of the book, when it became clear that I was learning more about history than about science. Only the last three of the 13 chapters come anywhere near the scientific advances of the last 20 years, which is the time when exploration of the middle world began in earnest.

Of these last three chapters, there is one about how Nature harnesses Brownian motion to drive (rather than hinder) the movements of molecular machines, but very frustratingly, we aren't really told how this works. For example the Brownian ratchet mechanism is only hinted at, but not mentioned explicitly. Then there is one chapter about "middle world technology," otherwise known as nanotechnology, where the author again manages to escape to Ancient Egypt while telling us only a little bit about the present time. The final chapter offers the suggestion that life could only arise because of the peculiarities of the middle world.

All in all, the book leaves the impression of a very beautifully crafted Trojan horse that could have smuggled quite a lot of science into the walled cities of the non-scientists. Sadly, the author failed to include most of the payload. Between all those anecdotes and excursions into history, one could have easily concealed 100 pages worth of scientific discoveries from the last 15 years, including reasonably detailed accounts of molecular machinery of the cell and nanotechnology. A few illustrations would also have been useful.

Ultimately, the satisfaction you can get out of this book will depend on what you expect. Readers who yearn for fresh excitement from the frontiers of research may find it diasppointing. However, if you want to indulge in the pleasure of a rollercoaster ride around the history, philosophy, and celebrity gossip of an unfairly neglected part of science, the middle might end up being just right for you.

The world according to Bart

What's science ever done for us?
Paul Halpern
Wiley 2007

Originally published in Chemistry & Industry, No.2, (28.1.2008) 29.

Popular culture and science don't normally fit together very well. From Frankenstein to Strangelove, there is a long tradition of depicting scientists as deranged or plain evil. If today's TV culture is any indication of the role that science has to play in society, there are only two situations which require the services of a scientist. You need a boffin, firstly, if you want to travel around the universe and/or communicate with aliens, and secondly, if you have been murdered, typically by a serial killer. Happens to the best of us. In the latter case, of course, your personal benefit is only limited.

Beyond these two scenarios, the most-used medium doesn't offer much of a glimpse of what science has actually given us, from the TV set itself, through to today's iPods and flash drives. Not to mention that half the nitrogen in our bodies has been through the Haber-Bosch process. Enlightening TV formats like the BBC's Tomorrow's World have given way to stunt shows clearly aimed at braindead males with testosterone poisoning. If the demonstration of Newton's laws of gravity by "human guinea pigs" jumping into a swimming pool is as much science as TV can offer, we are in deep trouble.

Fortunately though, we still have The Simpsons, as the last refuge of science and some residual intelligence in today's world, to answer barkeeper Moe Szyslak's pertinent question that provided the title of this book. Around the Springfield nuclear power station, a broad spectrum of sciences from psychology through to mathematics is mined for the creation of storylines and jokes.

In fact each of the episodes could be described as a scientific experiment. The whole of Springfield, obviously designed as a model system for society at large, is in a finely balanced equilibrium, where nothing changes in the long term. At the beginning of each episode, a small disturbance triggers a reaction which throws the whole system off balance, but in the course of less than 30 minutes, the reactions, which can lead to some spectacular fireworks along the way, will have found the way back to the initial equilibrium.

The notable exception to that rule is the series of Halloween episodes, known as "Tree House of Horror", where the disturbance tends to spiral out of control and lead to a more or less catastrophic ending. But rest assured that by the beginning of the next episode, the initial equilibrium is back where it was, regardless of whatever disaster befell our yellow-tinged friends over Halloween.

Within that experimental framework, there is obvious scope to experiment with (almost) human subjects without having to seek approval from an ethics board, so all the main characters suffer some experimental change of their lives at some point. Beyond the psychology, education, and society aspects, however, there are also lots of experiments – and one-liners – from the natural sciences.

There is of course the nuclear power station, always on the brink of a disaster, not least because of the work of its least qualified employee, Homer Simpson. Radiation leaks have sped up evolution and given the world Blinky, the three-eyed fish. A separate incident involving radioactivity leads to the tomacco, a tomato with the addictive properties of tobacco.

References to chemistry range from biochemistry (e.g. Lisa's groundbreaking discovery of the pheromone that drives bullies to attack nerds) through to the laws of thermodynamics, which Homer tries to enforce in his household. Even though he claims, when playing Scrabble, that nobody could possibly form a word from the letters: O, X, I, D, I, Z, E. The physics of the Simpsons includes applied engineering (magnetically levitating monorail trains) and the purely curiosity driven, such as the investigation of the Coriolis effect and the rotation of water draining from an Australian sink.

Paul Halpern, a professor of physics at the University of the Sciences in Philadelphia, has now published the first book on the science of The Simpsons. He follows a modern tradition that started with Lawrence Krauss's The Physics of Star Trek, but the obvious difference from this and many other "Science of" titles is that its subject is not located in science fiction or fantasy. The Simpsons are based on modern life, and the main deviation from real life is that the characters are yellow (or otherwise implausibly coloured) cartoons with only four fingers on each hand.

Thus, the fact that the series considers science as a significant part of the modern life it chooses to lampoon must be applauded, and a book exploring this new interface between science and popular culture can offer much more than just a nitpicker's guide to some trivial TV entertainment.

Halpern offers a guided tour through the science of The Simpsons in 26 short and entertaining chapters, organised in four parts, which cover biology, mechanics, time, and the universe. The author, who has also written books on astrophysics and cosmology, is clearly more comfortable with the topics closer to his expertise, which means that the book improves as he moves on from biology to the topics he is more at home with.

The best chapters in the more physical parts of the book are perfect vignettes of important scientific topics reflected in one of the most intelligent products of modern popular culture. The book is also a valuable resource for scientifically minded Simpsons fans to look up science references that they didn't quite get when watching the show (yes they are that clever!) and to put the ideas hinted at in the series into a broader scientific context. Last but not least, it is a valiant attempt to smuggle some science into the heads of some readers who might otherwise never have picked up a book about a scientific subject.

However, the book is unlikely to remain the last word on the science of this series, as, firstly, the series is still going strong and producing more scientific results all the time, and secondly, the author's focus on physics and cosmology leaves space for a more detailed analysis of biology (did anybody get the botany joke written in the flower beds outside the botanic garden, which Lisa offered to explain, but then didn't as nobody was interested?), chemistry, mathematics … In fact, there is a whole website dedicated to the mathematics of The Simpsons, simpsonsmath.com, with details of mathematical references and the crucial piece of information that Simpsons producer and writer Al Jean holds a mathematics degree from Harvard.

So watch this space … and The Simpsons. Scientists have proven that there is nothing quite like it on the box.

Right answers for wrong questions

10 questions science can't answer (yet)
Michael Hanlon
Macmillan 2007

Originally published in Chemistry & Industry, No.13, 7.7.2008, 28.

The most important skill in science, as I remember several of my research supervisors saying, is to ask the right questions. To get ahead in modern-day science, you need to find questions that are important, of course, either for fundamental understanding of our world, or for practical applications, or even both. Among all the important questions you can think of you then need to pick those that cannot be answered today, but which hopefully, with a bit of luck and stamina, you may be able to answer tomorrow.

This concept of "tomorrow," though not meant literally, has shrunk significantly over the last few decades: From the thirty-odd years for the first crystal structure of a protein (myoglobin, published 50 years ago), via the decade for the human genome, we have now arrived at the point where funders and employers will insist on getting the answer within two years. Accordingly, it is a real challenge to find an important problem that cannot be solved today may yield to a research effort of just two years.

Michael Hanlon is the science editor of a tabloid newspaper and not a research scientist, so he is free to throw up whatever questions he likes, without worrying whether they can be answered within two years or even within two hundred years. He picked ten questions that science can't answer (yet) and wrote a 20-page chapter about each of them. But has he picked good questions, maybe the kind of question that can drive research in the right direction, even if the goal remains out of reach?

His list is a very mixed bag. In my opinion, there are only two genuine scientific questions, of the kind which we should be able to solve but haven't quite got the ability to, namely the nature of 96 % of the Universe (those parts that we call dark matter and dark energy, to disguise the embarrassing fact that we haven't got a clue about them), and the question where and how life originated and how common it is in the Universe.

Then there are four questions from the science/philosophy interface. While there are scientists (and philosophers) dealing with these questions, I happen to believe that they are not suitable to be addressed with our current scientific toolkit, essentially because they lack the screws and fixtures that our tools would fit into. These questions are about the nature of time, consciousness, reality, and personal identity.

Finally, there are four questions which in my view have very little to do with science, as they are about people and public policy. Essentially, they are about people who are too dim to succeed in today's highly technical society, those who are too fat, those who believe any old nonsense, and those who want to live forever. Before reading the answers, I would have dismissed these questions as irrelevant to science, but afterwards I kind of agreed that there was at least a little bit of a scientific angle to be gained from two or three of them.

So how about the answers? Quite surprisingly, given the uneven quality of the questions, Hanlon's responses to them make quite a lot of sense most of the time. For the two proper science questions, he gives concise and accessible summaries of these seriously difficult, if not intractable, issues.

On the four philosophical questions, the shortness of the chapters is their saving grace. While I might not have the patience to sit through an entire book about the nature of time or consciousness, his short treatments are again readable and manage to give an impression of why these questions are so hard and possibly outside the range of what science can answer within reasonable time.

The four policy questions, finally, were treated much more sensibly than I might have expected from somebody who works for a tabloid newspaper. I found myself agreeing with much of the content, even if I didn't find all that much that pointed to possible future research projects. The one point where I did learn something that was both new to me and potentially a starting point for actual scientific research was the question as to whether the current obesity epidemic might have biological causes beyond the trivial ones of too much food and too little exercise. He very nearly convinced me that the wild hypothesis of an infectious agent making people more susceptible to obesity may be worth looking at scientifically.

So, all in all, Hanlon has managed to write a nice little book about ten questions, eight of which I as a scientist wouldn't have accepted as meaningful scientific questions beforehand. And he may even have won me over on one or two of those eight.

Now if somebody could do the same for ten real scientific questions, that would be even more interesting (and quite possibly less successful commercially). The trouble is, however, that anybody who is able to come up with ten really good scientific questions is probably too busy writing grant proposals and bossing an army of post-docs around, to even consider writing a popular science book.

Microbe hunters become the hunted

Deadly companions: how microbes shaped our history
Dorothy H. Crawford
Oxford University Press 2007

Originally published in Chemistry & Industry, No.15, 11.8.2008, 29.

In 1926, Paul De Kruif wrote the popular science classic "Microbe hunters", setting the optimistic mood that was to prevail throughout much of the 20th century, when scientist believed they could win, and were winning, the fight against infectious diseases. Shortly afterwards, Fleming discovered penicillin, which was followed by many other antibiotics. Simultaneously, new vaccines were developed and helped to ban virtually all deadly infections from the wealthier nations of the earth. When smallpox was eradicated from the wild in 1980, the general expectation was that other diseases would swiftly follow it into oblivion.

Today, however, smallpox is still the only disease that we have wiped out completely. In sharp contrast to the hopes and expectations of 1980, infectious diseases like malaria, AIDS, tuberculosis and many more are now killing more people than ever, and there are very serious concerns that the globalised human society may be catastrophically vulnerable to new pathogens which appear to arise at a rate of about one per year. So what went wrong, when did the microbe hunters become the hunted?

Dorothy Crawford, a professor of medical microbiology at the University of Edinburgh, has taken a sabbatical year to explore our history from this angle – an all too rare example of an established scientist taking time to popularise science. Her book is an excellent explanation of what went wrong for most of human history, what went very briefly right in the 20th century, and why microbes have now wrongfooted us again, at the beginning of the 21st. As microbes are everywhere around us, most of this "deadly companionship" can be explained in terms of human activities and lifestyle changes offering pathogens from other animals new opportunities to invade.

Thus, Crawford very convincingly outlines how the beginning of agriculture and the first cities opened the doors for zoonoses (diseases that crossed over from animals) that were unknown to our hunter-gatherer ancestors. Critical population sizes and timescales of travel determined which diseases could establish themselves in the human population, and which could only flare up and burn out. Those that established themselves, like measles, typically became less virulent over time, as they co-evolved with their human hosts, becoming the "deadly companions" of the book title.

Looking at history with an epidemiological perspective, much of the progress of the past centuries (bigger cities, international trade, faster travel) was good news for pathogens and very bad news for large parts of the human population. Perpetual warfare also encouraged the spread of infections. And for most of our history, medicine wasn't helping much either. In the case of smallpox, Crawford tells us, it was noticed in the seventeenth century that the rich people who could afford treatment were more likely to die of the disease than the poor who couldn't.

Humans went into the offensive in the brief period covered by "Microbe hunters", beginning with the discovery of microbes by Antoni van Leeuwenhoek. Then, the age of antibiotics came along and seemed to herald our victory over infectious diseases. However, in ways that were only fully appreciated in the 1990s, antibiotics were their own undoing, as they helped to spread resistance genes that had been rare among bacteria (but which existed, because antibiotics are an old weapon from the longstanding competition between bacteria and fungi). As a consequence, we now have MRSA "superbugs", along with drug resistant strains of diseases like tuberculosis and malaria.

This crisis has been accelerated by the indiscriminate and irresponsible use of antibiotics, a global self-inflicted problem that is still underappreciated. In countries where antibiotics are available over the counter, people may end up using them when they are not needed or may fail to finish an antibiotic course, thus spreading resistance. In environmental microbes and animal pathogens, the widespread use of antibiotics in agriculture is also helping to breed resistant strains. More than half the global production of antibiotics, Crawford reminds us, ends up in agriculture.

Eventually, it may turn out that this half is endangering as many human lives as the other half saves.

One silver lining in the "superbug" crisis which Crawford does not mention, is in current efforts to develop alternatives to antibiotics. Researchers interested in disrupting the synthesis of the bacterial "hairs" (pili) which help many pathogens to stick to their hosts, hope that the disruption of crucial virulence factors without actually killing the microbes will be less likely to breed resistance, and also give the immune system a better chance to develop an immunological memory of the pathogen.

And then there remains the growing problem of viruses, which are beyond the powers of antibiotics. The rise of AIDS in the 1980s was probably the first major hint indicating that infectious diseases weren't going away. Since then, new pathogens, many of them viruses, have emerged at a rate of about one per year, including Ebola, SARS, and avian influenza H5N1.

As Crawford explains quite lucidly, these new killers are very similar to the very old ones, like smallpox and the bubonic plague, in that they emerge from animal reservoirs to hitch a ride on new(ish) human lifestyles. The combination of traditional closeness between men and beasts in China gave rise to SARS, but international air travel was the key to its global spread after it had evolved the ability to propagate among human patients. Without blanket antiviral drugs that could protect us from newly emerging virus strains, we may soon be as exposed to the threat of infectious disease as our ancestors were before antibiotics were invented. After a brief period of posing as highly successful microbe hunters, we have become the hunted again.

Celebrate our luck

The Drunkard's Walk: How randomness rules our lives

Leonard Mlodinow

Allen Lane 2008

Originally published in Chemistry & Industry, No 22, (24.11.2008), 31.

We are all extremely lucky to be here. If you consider the odds against a particular sperm fertilising a particular egg and establishing a successful pregnancy, every person alive is a winner of the biggest lottery in the Universe.

Yet we don't generally appreciate the important role that randomness plays in our lives from the very beginning. And we often misjudge probabilities, a problem that can have catastrophic consequences every time when a statistical argument is used in a court of law or to gauge the safety of a given activity.

With this book, physicist and screenwriter (what are the odds against that combination?) Leonard Mlodinow is trying to educate his readers about elementary statistics, using colourful anecdotes from the dawn of this science, and to make them realise how widespread misconceptions of probabilities and randomness are in our society.

For me, the highlight of his book is the short biography of the gambler, medic, inventor, and arguably father of statistics, Gerolamo Cardano (1501-1576). He published 131 books, invented a part that you will find in every car on the roads today, and made a living from applying statistical analysis to gambling at a time when everybody else considered the outcome of chance events as determined by the will of God.

With the spellbinding lives of Cardano and subsequent luminaries including Pascal (with his famous triangle), the Bernoulli clan, and Thomas Bayes, Mlodinow manages to make the normally boring science of statistics an interesting read. Not to become bogged down in history, he intersperses these parts with many examples of misuse of statistics and probability in modern life.

His main mission appears to be the fight against our failure to appreciate that success in life may in many cases be due to pure random luck. Especially in areas like music, sports, and the film industry, where many talented people compete but only very few become superstars, we tend to interpret a streak of luck (like 5 wins in a row against an equally good opponent), as a sign of towering talent. Mlodinow points out again and again that even when flipping a fair coin many times, one will at times see streaks of heads, purely due to the randomness of the process. Thus, he argues, if between two comparable teams or athletes, one wins five times in a row, this could still be random chance.

Putting a positive spin on this roulette wheel of life, however, Mlodinow reminds us that everybody can try their luck as often as they wish. Often, the people who eventually do hit the jackpot, have simply thrown their dice more often than others.

While the thin line between luck and talent in sports, music or film matters mainly to the participants and to the aficionados, it can have major consequences in the world of finance, where fund managers are paid astronomic salaries for picking stocks to invest in. As thousands of people engage in this activity, Mlodinow argues, it is inevitable that some of them will have streaks of good luck that may even last for many years and create the impression that they have a "golden hand" in picking the right investment.

Crucially, however, if one tries to use the success of fund managers in the past five years to predict their success in the next five years, it will all dissolve in a pattern that looks suspiciously like random noise. In other words, if all those superstar fund managers made their decisions by flipping coins, the result would be indistinguishable from their current activity. Or should we conclude that they are flipping coins already?

Still, we can opt out of this kind of misuse of mathematics by not investing in funds managed by overpaid coin-tossers. Some people, however, find themselves the helpless victims of mathematical ignorance when they appear in court and are told, for example, that it was extremely unlikely that both their babies died by accident, *ergo* they must have murdered them. This is arguably the area where Mlodinow's message is most desperately needed. In such cases ignorance still ruins lives, as the author demonstrates convincingly using the examples of several prominent court cases.

All in all the book is a compelling read and points out very important flaws in the ways in which we as a society, and our decision makers in particular, deal with statistics and randomness. With his focus on fund managers and baseball players, he only gives little coverage to randomness in Nature, such as Brownian motion. Also, he completely ignores the idea of multiple universes, a weird but scientifically respectable way of accounting for the random behaviour of quantum mechanics.

Combined with his assertion that many of the top earners in our world owe their good fortunes to sheer luck and possibly a stubborn persistence in trying their luck again and again, it would be rather amusing to speculate on the fate of alternative universes where Madonna serves tables in Michigan and George W. Bush remained an un-reborn alcoholic for the rest of his life.

Regarding the randomness at the start of his own life, Mlodinow hints at how the Holocaust, by wiping out his father's first family, led to his own existence, but he doesn't explore the randomness around conception, which is arguably one of the largest sources of randomness in our world. If, for instance, Hitler had failed to grasp power in 1933, the lives of almost everybody living in Europe would have taken different pathways from what actually happened. Because of that, people would have had different sets of children, so practically nobody born in our world after 1933 would have been born in that alternative universe, and Europe would today be populated by a completely different set of individuals. Leonard Mlodinow wouldn't exist, nor would his book, nor would I be at hand to review it.

Thus, we are essentially random people in a world governed by random molecular motions. We should celebrate how lucky we are to be here against all odds.

Life's enduring mystery

Origin of life: Chemical approach
Piet Herdewijn, M. Volkan Kisakürek (eds.)
Verlag Helvetica Chimica Acta / Wiley-VCH 2008

Originally published in Chemistry & Industry, No 12, (22.06.2009), 31.

The year 1953 saw the publication of two major breakthroughs in biology. On April 25th, Watson and Crick published their double helix model in Nature. And just weeks later, Stanley Miller reported in Science magazine that amino acids can arise from simple inorganic chemicals in a system modelling the conditions thought to have dominated in the early days of our planet. Thus the molecular basis of how life works and how it came into existence seemed to be within reach. Optimistically inclined observers may have thought at that point that the rest was just a question of filling in the details.

In the 55 years that followed, molecular biology has delivered an amazing amount of details concerning how life works, culminating in the current flood of genome sequences of many different organisms. It has now reached the point where the details can be integrated into an emerging understanding of how genomes function at a system level. Everybody has at least a vague idea of what kind of knowledge grew from the seed of the DNA structure. But what came out of Miller's simultaneous, equally fundamental origin of life experiment?

Obviously, the mechanism of current life is easier to study than its origins, which happened more than four billion years ago, when no one was watching. Thus, it is not surprising that much less progress was made in Miller's field than in Watson and Crick's. And yet, there are a few good reasons why we might be able to work out what happened all those years ago.

In a first step back in time, we know that all organisms on the planet today descend from a common ancestor, a microbe known as LUCA, for Last Universal Cellular

Ancestor. With the growing knowledge of the genomes of a large number of species alive today, we should be able to reconstruct LUCA's genome and find out a lot more about this ancestor than we know today.

But this is only half the distance. LUCA was already quite evolved in that it had DNA, RNA, and proteins including lots of different enzymes. Connecting LUCA to the inorganic chemicals in Miller's apparatus is the most challenging part of the problem.

There are a few further clues to be found in today's biochemistry. For instance, the ridiculously complex machinery that makes proteins based on the genetic instructions read from messenger RNA is older than LUCA and contains records of prior evolution. The enzymes that bind amino acids to transfer RNAs are grouped into families, reflecting the fact that they evolved (before LUCA's time) from only two or three different proteins. Similarly, the tRNAs themselves evolved from a smaller number of adapter RNAs, and the ribosome must have started out as a catalytic RNA in a world without proteins, the so-called RNA world.

Genetics deserts us in the final part of the quest, connecting the RNA world to the chemical origins. Present-day clues for this part of the puzzle are mostly found in bioinorganic chemistry. For instance, today's biochemistry uses some metal ions that aren't particularly common or easy to handle, such as molybdenum, and copper. Does that tell us something about the precursors to today's enzymes? Did inorganic catalysis precede enzyme catalysis?

Trying to bridge this gap from the other end, researchers are also thinking about ways of violating Pasteur's treasured rule that life doesn't spontaneously arise from inanimate matter. At least once, a long time ago, it must have done just that, and chemists are still trying to figure out how that happened.

This book mainly deals with this earliest and most challenging stage in the history of life on our planet, the path from the chemical origins to the RNA world. It is a collection of papers picked from the journal *Chemistry and Biodiversity*, dedicated to the memory of two pioneers of the field, Stanley Miller and Leslie Orgel, who both died in 2007.

The papers include reviews and original research papers, and they don't come with any additional explanations or presentation, so the book doesn't offer easier access to the field than an academic journal or the proceedings of a conference would. Its main function is as a repository of the latest thinking in this field by some of the key researchers (including Eschenmoser, de Duve, and Wächtershäuser) at the moment defined by the exit of Miller and Orgel.

The prevailing impression is that of frustration. Fifty-five years after Miller's key experiment, there are many more details known about the parameters constraining the mechanisms by which life on Earth originated, and the book gives an impression of those details. But an understanding of how these pieces might fit together is still so far away that we can't even tell whether it will ever arrive.

My own gut feeling is that the necessary information is there in the genetic, chemical, and geological data, but we (i.e. all of humankind together) just aren't clever enough to put it together to reveal the correct and definitive picture. Maybe the most original and influential thinker on the origin of life over the last couple of decades is Günter Wächtershäuser, a patent attorney who came to the field as a hobby researcher drawing up new theories in his spare time. His chapter in the book is definitely worth reading, even if it leaves the reader with the frustrating thought that if Wächtershäuser can't crack the conundrum, nobody can.

For better, for worse

The alchemy of air: A Jewish genius, a doomed tycoon, and the scientific discovery that fed the world but fueled the rise of Hitler
Thomas Hager
Harmony books, New York, 2008

Originally published in Chemistry & Industry, No 7, (13.04.2009), 30 .

Turning nitrogen from the air into fertiliser is by far the most important process that the chemical industry is carrying out today. The rule of thumb is that the high pressure synthesis of ammonia invented by Fritz Haber and developed to industrial applicability by Carl Bosch turns over a comparable amount of nitrogen as the natural version of the process, nitrogen fixation by bacteria. Thus, half the nitrogen that we eat and half the Ns in our proteins and nucleic acids are bound to be the product of a Haber-Bosch factory. This process literally keeps us alive.

However, the process has also become a prominent example of dual use science. It has helped to feed the world, but it has also helped to provide the explosives for both World Wars. Moreover, both Haber and Bosch have done significant harm that has prevented them from being celebrated as heroes. In Haber's case it was the development of chemical warfare during World War I. And Bosch, though far from sympathising with Nazi ideology, obsessively pursued the idea of converting coal to petrol, which ended up fuelling Hitler's tanks and planes as they conquered Europe.

The development of the Haber-Bosch process essentially shaped the 20th century, for better, or for worse. It is a frightening idea that the history of the century could actually have turned out even worse without this invention than it did with it. Global famine and wars over food and fertile land fought with sabres, gas, and nuclear weapons (remember, we're short of explosives!) could have made for a very different, and probably not more agreeable 20th century. Note that this alternative isn't that far out, as the odds against Bosch's success with the industrial process were enormous.

Considering all this, surprisingly little has been written about Haber and Bosch. It took until 1994 for the monumental and ultimately tragic life of Fritz Haber to be tamed and locked up between book covers. The German original of Dietrich Stoltzenberg's biography of Haber, based on materials collected by Haber's assistant Johannes Jaenicke, is a hefty tome of 685 pages, but the English translation has been shortened somewhat. As far as I am aware, there is no comparable biography of Carl Bosch.

It is difficult for writers to find the right way to approach this monumental complex of tragic lives, science and technology of the highest possible impact, and politics gone off the rails. Apart from the epic biography of Haber, I have on my shelves a few slimmer studies of smaller aspects. There is a biography of Haber's first wife, Clara Immerwahr, for instance, and a master's thesis on Haber's quest to mine gold from the oceans. Plus a 1950's account of the history of IG Farben dressed up as a novel, and a volume with short biographies of Jewish scientists in Germany 1900-33.

Now a fresh generation of American writers seems to have discovered this topic for themselves, and they approach it fearlessly. A few years ago, there was a new, shorter biography of Haber by Daniel Charles, who delivered a straight and very readable narrative of Haber's ambivalent character and his rise and fall.

Thomas Hager has taken a step back to glance at an even bigger picture, including the history of nitrogen fertiliser before it could be made synthetically, and the further development of Bosch's ambitions after the success with nitrogen. Essentially, he follows the intertwined stories of three protagonists: Haber, Bosch, and nitrogen.

As a somewhat exotic prelude, Hager offers us a brief history of the 19th century trade with nitrogen fertilisers, first with guano from the rocky islands off the Peruvian coast and then with saltpetre from the Atacama desert, which became part of Chile in a war fought over nitrate. Part two describes the long road from Haber's first desktop reactor spitting out millilitres of ammonia, to the large scale plants producing tons of it, with all the drawbacks, pitfalls, and challenges on the way. Most chemists reading this book will probably be so familiar with the end result of the development that they feel the urge to shout at the characters in the book to help them along. But Bosch and his coworkers did get it right in the end, and the whole painful development process serves as a reminder of how easily the whole enterprise could have failed.

And then they lived happily ever after … or maybe not. The final part of the book bears the cryptic title "syn" which is presumably a merger of sin and synthetic fuel. First Haber sinned by developing chemical weapons in the erroneous belief that they would shorten the war and thus reduce suffering, and of course let Germany win. There was a deeper motivation as well: as a converted Jew, Haber was desperate to shed the image of a second-class citizen and become a proper member of the German elite. Helping to win the war might have been his ticket, but it didn't work out. For similar reasons, he later embarked on the mildly eccentric quest for the gold of the oceans, which he hoped would help Germany pay the reparation debts.

Then Bosch, who had risen to the position of CEO at BASF and then at the conglomerate IG Farben (formed of BASF, Bayer, and Hoechst), became obsessed with the development of synthetic fuel made from coal. At first, it appeared that oil was running out, in which case fuel from coal would have found a huge market. Bosch believed fuel would be the next big thing after nitrogen fertiliser. Then came the discovery of the Oklahoma oil fields and synthetic fuel was no longer competitive. Bosch pressed on regardless, and obsessively, throwing good money after bad.

While synthetic fuel didn't make sense for the global market, it made a lot of sense for the war machine that Hitler started to build up as soon as he seized power. After all, you can't depend on oil imports when you're at war with the rest of the world. Although he didn't approve of Nazi politics and was quite happy to say so, Bosch led IG Farben into a close association with the Hitler regime, simply on the grounds that the regime would happily buy as much fuel as Farben could produce. Blinded by his obsession with synthetic fuels, Bosch realised too late that he effectively fuelled the next World War.

Hager tells this story with verve and particular enthusiasm for action, including battle scenes, explosions and such like. Covering so much ground in just 280 pages, the characterisation of the protagonists has to remain somewhat superficial, compared to proper biographies. But they are three-dimensional enough to make the story work, and to make it interesting even for lay readers. In the attempt to attract this ever-elusive audience, Hager has kept all chemistry out of the text, referring to chemicals only by common names such as Chilean saltpetre for sodium nitrate. And the publishers helped him out with a sensationalist title clearly aimed at the readers of the tabloid press.

Let us hope that the lay readers do pick up his book, as it tells the most important story of the last century.

Reasons for asymmetry

The origin of chirality in the molecules of life
Albert Guijarro and Miguel Yus
RSC Publishing 2009

Originally published in Chemistry & Industry, No 9, (11.05.2009), 30 .

The genius of Louis Pasteur (1822-1895) is often exemplified with two experiments that defined his scientific legacy. First, in 1848, still early in his career, he picked apart the two types of crystals of "racemic acid" and showed that this optically inactive substance was a mixture of the enantiomers of tartaric acid. For the first time, he linked the macroscopic asymmetry of the crystals to a molecular property, the rotation of light.

In 1861, he published detailed work which summarily dismissed the then-popular idea of "spontaneous generation", i.e. the emergence of living organisms from dead matter. Using sterilized materials and the famous swan-necked flasks designed to avoid contamination by germs dropping in from the air, Pasteur showed that life cannot arise out of non-living materials.

Both these experiments still find echoes in today's science, because on at least one occasion in the distant past, nature must have violated Pasteur's dictum and produced life out of inanimate matter. Details of the origin of life have eluded researchers to this day. And when it happened, nature also picked apart the left and right-handed forms of the molecules involved, which is why virtually all biomolecules are chiral to this day.

One of the less publicized aspects of Pasteur's life is the fact that he spent a lot of time and effort looking for ways in which nature could have performed this original symmetry breaking, and didn't find any. By now, however, researchers have come up with a few processes that could in principle achieve such a feat. The quest for the origin of chirality has emerged as a potentially manageable part of the more daunting problem of the origin of life.

Albert Guijarro and Miguel Yus describe the chirality problem and our current best guesses at its solution very systematically in their book. They start from first principles with the physics of time and space and precise physical definitions of symmetry and asymmetry. They very usefully split the problem into two fundamental spects: symmetry-breaking events, and mechanisms that could amplify existing imbalance between enantiomers.

At the fundamental physics level, it is clear that of the fundamental particles and forces of the standard model, the weak interaction is the only one that can create chirality, but whether and how it can give rise to chirality in biomolecules is still unknown.

In terms of practical laboratory experiments that can break symmetry, researchers have only recently found the examples that eluded Pasteur. In 2001, J. M. Ribo and coworkers succeeded in creating chiral supramolecular aggregates with the help of the chiral swirl of the liquid in an ordinary rotary evaporator.

However, for the origin of life to result in an enantiomerically pure biomolecular world, a fundamental symmetry-breaking event may not even be necessary. If life only arose at a small site, a small random local imbalance may have been amplified by autocatalytical networks of chemical reactions. Such mechanisms were first described by F.C. Frank in 1953 and have since been demonstrated in practice as well. Experimental implementations of such autocatalytic systems include organic syntheses reported by K. Soai in 1990 and the self-replicating peptides developed by M. Reza Ghadiri in 1996.

Thus, given the power of chemical reactions to select their reactants, we may not have to look at fundamental physics or the wider universe to understand why life on Earth uses only D sugars and L amino acids. Still, finding other examples of life in the universe would help us to understand whether this is the consequence of symmetry breaking or of the amplification of random fluctuations.

Although they are affiliated with the department of organic chemistry at the University of Alicante, Spain, the authors treat this topic with the merciless thoroughness of a theoretical physicist, as they slowly progress from the fundamental equations to the practical experiments. Cross-referencing is used in abundance, and physical equations outnumber chemical ones.

For the general reader interested in practicalities, the emphasis on physics and theory makes this a challenging read. Mountains of theory have to be overcome before one arrives at actual chemical reactions. Intriguingly, a more practically-minded account of the same topic appeared almost simultaneously (Uwe Meierhenrich: Amino Acids and the Asymmetry of Life: Caught in the Act of Formation, Springer 2008), which at first glance appears to be more accessible to lay readers. The present book may be most suitable for physicists or physical chemists keen to explore astrobiology as an optional subject or research sideline.

If the book can attract a few bright students to this intriguing field, it will have been successful. As with the wider investigations concerning the origin of life, we have barely begun to understand the questions and are still a long way from finding definitive answers or developing an understanding of how our strangely asymmetrical living world came into being.

Teaching DNA new tricks

Aptamers in bioanalysis
Marco Mascini, ed.
Wiley 2009

Originally published in Chemistry & Industry, No 17, (14.09.2009), 27.

For somebody coming from the world of protein research, where every molecule is an individual, DNA appears boring at first glance. With only four bases to play with, and the dogmatic double helix structure holding forth irrespective of the sequence, there appears little scope for imagination. Reading out DNA sequences for genome analysis is so tedious that we leave it to robots, and the excitement only sets in when one gets to analyse the data in depth and can draw comparisons at the genome level across the phylogenetic divisions.

Similarly, the method known as SELEX (systematic evolution of ligands by exponential enrichment), which involves the production of millions of random DNA sequences in the hope that some of them will do what one wants them to do, struck me as an uninspiring research task when I first heard of it. But here, as with genome sequencing, the assiduous use of boring work has eventually led to very interesting and useful results, in this case to the production of aptamers.

Aptamers are DNA or RNA polymers (or oligomers) with an aptitude to bind to a specific target. Originally, researchers developed these new specific binding molecules as a stable and manageable alternative to antibodies. While the specificity and binding constants are comparable, DNA aptamers are smaller and more stable than antibodies. Moreover, as their production does not involve animals, aptamers can also be produced against highly toxic compounds. Thanks to the techniques developed for molecular biology, aptamers can be easily produced at large scale, modified, immobilised on microchips, and analysed.

There is a whole range of possible applications requiring high performance molecular recognition of the kind that aptamers can achieve, from drug targeting through to molecular information processing. In 2004, the US Food and Drug Administration approved the first aptamer-based drug, Macugen. It specifically binds to a protein involved in the formation of new blood vessels and is prescribed against age-related macular degeneration, a common cause of vision deterioration in old age.

In a recent research paper, aptamers were used to target liposomes carrying the anti-cancer drug cisplatin specifically to tumour (e.g. breast cancer) cells. If it reaches the clinic, such work bears the promise of cancer chemotherapy without the side effects.

One large field where aptamers have shown particular promise is in biosensors and analytics of biomolecules. For instance, Kevin Plaxco's group at the University of California at Santa Barbara has very successfully combined spectroelectrochemistry with DNA aptamers that adopt different conformations in the presence and absence of the target ligand to create highly sensitive, specific, and user-friendly biosensors for proteins (e.g. thrombin), small molecules (e.g. cocaine) and specific DNA sequences.

DNA aptamers will typically be single DNA molecules not matched up with a complementary counter-strand in a double helix. Some parts of the sequence may be self-complementary such that they form hairpin structures or stems carrying loops. For biosensor experiments, one end of the DNA molecule will be immobilised to a solid surface, typically a gold-coated electrode. In the ideal case, known as the signal-on architecture, an electrochemical probe such as ferrocene is linked to a site that is far from the electrode, but will be able to come closer to the electrode after the aptamer has bound its target and rearranged its structure accordingly. Thus, the electrochemical signal will be off, while there is no ligand, and it will be switched on when the ligand binds to the aptamer.

Unlike most other biosensor technologies, this approach does not require chemical additives and can be used even with contaminated samples and crude solutions such as unmodified blood. All that the user has to do is to dip the sensor into the liquid to be analysed, and read the answer from the electronic display.

"Aptamers in Bioanalysis" is to my knowledge the first book dedicated to these exciting new developments, so I hoped it might provide not just an overview of the field but also an introduction, as much of the material will be new to most people. It comprises 12 chapters by different authors or teams of authors, including two introductory chapters on aptamers and SELEX, six on aptamer biosensors, and four on analytical applications.

The chapters appear thorough as scholarly overviews of the relevant fields, with instructive diagrams and very comprehensive references. What the book doesn't achieve, however, is to communicate the importance of and the sheer excitement surrounding a new, rapidly growing field that has enormous promise in applications from medical diagnosis through to environmental analytics and security.

I have a nagging suspicion that the intrinsically tedious nature of working with millions of nearly identical DNA molecules has dominated the writing of the book, where the enthusiasm over new departures and discoveries should have prevailed.

Thus, while this book is perfectly acceptable as a scholarly reference work, there remains an opening for an introductory work to attract students to the field and to communicate its relevance to a wider audience.

Psychic papers

Plastic fantastic: how the biggest fraud in physics shook the scientific world
Eugenie Samuel Reich
Macmillan 2009

Originally published in Chemistry & Industry, No 19, (12.10.2009), 26.

Fans of the British cult TV series "Doctor Who" may remember that the Doctor (the extraterrestrial lead character whose academic credentials are never explained) uses "psychic paper" – a card-like material that displays to the observer exactly what they want to see. This gimmick comes in handy whenever an ID is required to get into top security government offices and such like.

In the years 1998 to 2002, the doctor of natural sciences Jan Hendrik Schön (whose academic title was terrestrial and genuine but later revoked because of the events discussed below), created a real-life equivalent of psychic paper: a series of over 30 scientific papers that showed the readers exactly the results they had hoped for.

As Reich explains in her book, Schön perfected the art of "doing science backwards" i.e. figuring out which conclusion would be desirable, and then creating the matching datasets by editing existing experimental results. He appears to have been particularly good at sensing what his colleagues in theoretical and experimental physics hoped for, and then to produce these results his way, which he published in an unprecedented string of high-profile publications, including 15 in Nature and Science.

But how could he possibly get away with this for several years? Reich highlights several weak points that he skilfully exploited. Firstly, Bell Laboratories, where Schön worked as a post-doc from 1998, had a reputation as a world-leading laboratory, where the transistor had been invented in 1947, and where patents and papers were still being churned out at the time he arrived. Although by that time the labs began to feel the financial pinch and were far from their glory days, the address still opened doors for him and it appeared entirely plausible that a string of breakthrough discoveries should emerge from there.

While he used the Bell Labs address successfully, he withdrew himself from supervision there by frequently travelling to his native Germany and claiming he did key elements of his work there. Thus, whenever people wanted to see his samples or methods, he referred to the other lab, and only shared edited results. Similarly, the name of his supervisor Bertram Batlogg helped Schön to get his papers published, without Batlogg having much control over them.

The second weak point was the eagerness of the leading journals Nature and Science to attract the most spectacular breakthrough reports. At the time of Schön's miracle years, both Nature and Science were actively trying to get a larger proportion of physics papers on board, such as to broaden the scope of their readership and not to be seen as mainly biological journals.

While the referee reports that led to the acceptance of seven of Schön's papers by Nature and eight by Science are still kept under wraps, Reich claims that several referees were unhappy that technical criticism they had made during refereeing was not addressed in the published papers. From her contacts with people involved with these processes, Reich concludes that both journals paid more attention to the big impact of the conclusions (which of course were tailor-made to be exactly the kind of results that Nature and Science wanted!) than to the nerdy little details of the experiments. Ironically, some less prestigious journals might have paid more attention to the technical bits and thus might have ended the string of Schön's "discoveries" sooner, as he increasingly re-used the same datasets and made mistakes in using data that were incompatible with the experiments he claimed to have carried out.

So which lessons should be learned from the whole affair? Reich comes across all radical in her introduction, questioning whether the "self-correction" of the scientific record actually works, and why it worked so slowly in this instance, but at the end of the book she doesn't offer any definitive answer to that and leaves the readers to reach their own conclusions.

The weakness of Bell Labs is a unique feature of the Schön case, which can be explained with the historical greatness and slow economical decline of his host institution, but this particular set of conditions is unlikely to occur elsewhere often enough to worry us. Also, the way in which the institution set up an independent inquiry into the affair has been widely praised as the best way to deal with it.

What is worrying, however, is that the journals that are widely seen as publishing the most important scientific papers are so vulnerable to somebody who just holds a psychic paper under their noses and pretends to give them exactly what they want. Considering the fact that a few papers in Nature or Science (much fewer than Schön had, actually) are sufficient for a first author to secure a good job in a top university or company, these journals should take into account the possibility that people may cheat to attain this coveted prize.

While I haven't followed the fallout of the affair in Nature and Science in much detail, I didn't get the impression that they made revolutionary changes in response to the affair. One particular aspect the journals should take to heart is the way in which the referees' reviews, rather than detecting the fraud, helped Schön to improve his fabrications.

What is even more worrying is the fact that Schön might have very well succeeded in building a great career on fraud. What brought him down was the increasing number of errors and duplications that cropped up in his papers as he moved into more unfamiliar territory in the search of new areas to revolutionise. My suspicion is that he became addicted to the kick of having a breakthrough paper out and being praised for it, and as a result of that he acted less rational than he could have.

Had he acted in cold blood and considered his best path towards the Max-Planck directorship that he narrowly missed, he would have slowed down after six or seven papers in Nature and Science. He would have revolutionised fewer areas and backed up his first breakthroughs with some real research or at least with some more real-looking data. This way he would have committed fewer errors, his string of breakthroughs would have appeared less implausible and raised less suspicion. Critics would have taken a lot longer to figure out that there was something wrong with his work. And in that extra time, some of the breakthroughs that he anticipated might actually have happened, so he would have been vindicated by other researchers.

All this makes for a gripping story which Reich has researched and written up very competently. If I have one moan beyond the sloppy editing, it is that I really would have appreciated a list of Schön's publications in the appendix to find my way through all his breakthroughs. (I am counting myself extremely lucky that I haven't reported any of them, so I don't have to retract any of my articles because of him.)

For me the main lesson from the whole affair can be summarised thus: Never trust anybody who tells you what you want to hear. Especially not if they stand to gain substantial benefit from your attention. Not even if they are a doctor.

Going solar

The solar century
Jeremy Leggett (ed.)
Profile Books 2009

Originally published in Chemistry & Industry, No 20 (26.10.2009), 29.

Climate change has become one of the biggest issues of this century so far, and with it the need to switch our energy consumption from fossil fuels to renewable sources. One such source whose theoretical availability exceeds global energy needs by orders of magnitude is solar energy. Only a fraction of the world's desert areas would have to be covered with solar cells to provide for the entire world's demand. It is just a question of readjusting the economic framework (partially distorted by state subsidies for old and dirty technology) to make solar energy competitive in a free energy market. If we get that right, the 21st century could indeed become the solar century, and the threats of climate change and energy insecurity could still be averted.

How can it be done? One recent success story is the introduction of Feed-In Tariffs (FITs) in Germany in April 2000 by Gerhard Schröder's red-green coalition government. By guaranteeing that householders can sell back excess electricity from small scale installations to the energy providers, the government made the installation of solar panels (and other kinds of renewable sources) economically attractive. The subsequent boost in demand in turn allowed Germany's solar cell manufacturers to scale up production, which made the cells even more affordable, and started a virtuous cycle. By now Germany is not only one of the leading users of solar energy, but also one of the leading manufacturers of the corresponding hardware.

While this approach has something of a bottom-up, grassroots flavour about it, there are also some seriously big technology projects around that aim at providing much of our energy from large-scale facilities based in sun-spoilt areas. Most prominently, the DESERTEC project (www.desertec.org) involves installation of massive concentrating solar thermal generators in the North African desert and export of the electricity generated there via new direct current high voltage grids to Europe. This concept has recently won the official backing of a dozen major companies including Siemens, E.on, and Deutsche Bank, which joined up to form the Desertec Industrial Initiative, with the aim to develop concrete business plans and financing concepts for this large scale solar energy project.

Thus, one decade into what may become the solar century, significant progress has already been made in some places, and even bigger changes to our dealing with energy may still lie ahead. Considering that at the moment it is still cheaper to burn fossil fuel than to gain the same amount of energy directly from sunlight, the key challenge is to convince people that the switch to renewable energies must be done and that it can be done.

This is where Jeremy Leggett's book comes into the equation. It is a beautifully illustrated, well-argued plea to all who are still sceptical about renewables. In the first chapters, the authors explain why the dual crisis of energy insecurity and climate change makes a radical change necessary. Then they go on to explain how it can be done, in terms of light harvesting technologies, and in terms of investment and incentives needed to actually make it happen.

There are interesting explanations of how photovoltaic cells are manufactured today, and what factors determine the costs, but obviously, with such a slim and richly illustrated volume, one cannot expect every scientific question to be answered in detail.

The authors have no dogmatic preference for a specific type of technology and try to give all options fair coverage. Indeed, they emphasize that optimal use of renewable energies always requires a mixture of energy sources adapted to local conditions, and that solar energy should not be seen as the cure-all for all our worries. Mountainous countries like Austria and Norway already have fantastically high percentages of renewable energy thanks to the ready availability of hydroelectricity, while other areas may be more suitable for the exploitation of wind, geothermal energy, or solar.

Touchingly, the editor of the book claims to have installed the UK's first solar roof on his house in London, in 2000, i.e. at the same time as Germany introduced the FITs, which have since then been adopted in at least 37 countries. Now, the book says, there are hundreds of solar roofs in the UK, but there are hundreds of thousands in Germany, and home buyers there have the option to get a new house complete with the latest energy-saving technology and a solar roof installed at prices that don't exceed normal market prices.

Therein lies a very clear suggestion of what Her Majesty's government could and should do to boost its environmental credential and make at least a credible effort to meet the EU 20/20 targets. Alas, the solar century has had a rather sluggish start in some parts of the world, but it could still happen. And with the evidence for human-induced climate change piling up, it is becoming increasingly obvious that our century must become the solar century, or more widely defined, a renewable energy century, if it isn't to become a different book title, namely our final century.

Little giants

Giant molecules: from nylon to nanotubes

Walter Gratzer

OUP 2009

Originally published in Chemistry & Industry, No 2 (25.1.2010), 29.

Like ancient Gaul, chemistry is usually divided into three parts, inhabited by inorganic, organic, and physical chemists. As I don't want to get into trouble, I'll skip the bit about which of the tribes has the bravest warriors. Then there is that awkward no-man's land between chemistry and biology, which has been variously branded and rebranded as biochemistry, molecular biology, and now chemical biology. But some of the most interesting landscapes of chemistry actually stretch across two or more of these historical divisions, and suffer from not fitting into any of them. Among them are concepts like catalysis, and the field of macromolecular and polymer chemistry.

Polymers can be inorganic (e.g. poly molybdate complexes), organic (plastics) or biochemical (nucleic acids, proteins, carbohydrates), and all of these are studied by methods originating in physical chemistry. One could argue that they are an awkward topic, but one might just as well argue that the usual division of subdisciplines is awkward as it fails to accommodate a concept so essential for life and modern technology.

Life on Earth presumably started when something, somewhere managed to outwit entropy and produce ordered macromolecules from randomly floating monomers. Similarly, our modern technology really got going when people learned how to control the synthesis of large molecules. Without them we would probably live in a world more similar to those dreamed up by steampunk writers than to ours. And a real understanding of our own and everything else's biology depends on our ability to read (and also to recreate) biological macromolecules including DNA.

Deep down, every scientist knows that macromolecules are interesting and important, and that their collective science is underappreciated, so every once in a while, somebody tries to put things right by writing a book about how wonderful and important they are. I have one on my shelf which dates from 1952 and has been handed down the generations, and which has an equivalent title: Riesenmoleküle (this means exactly the same as giant molecules).

Now Imperial College emeritus Walter Gratzer has made his attempt to bring giant molecules to the masses. He recounts the short and chequered history of polymer science, with the fierce debates surrounding it. It always amazes me how long the scientific establishment resisted the very notion that proteins might be large molecules (as opposed to droplets of many small molecules). Might it have something to do with the well-established territories described above and the mistrust against any "foreign" concepts?

There are some cute factoids and anecdotes on the way. I wasn't aware, for instance, that JBS Haldane, anticipating the need for the DNA-untwirling enzymes we now call topoisomerases, said as early as 1953: "What you need is an untwiddlease." What a shame that his suggestion didn't make it into the official nomenclature.

Gratzer mainly dwells on how polymers were discovered or first produced (a lot of them by serendipity), their chemical structures, and what one can do with them. Sadly, he doesn't have as much time for the methods used to study them. For instance, he only very briefly mentions that in analytical ultracentrifugation you can look through a sample while it's being spun around at something like 60,000 rpm. I still remember how I saw this for the first time in one of the two wardrobe-sized ancient Beckman model E ultracentrifuges which my PhD supervisor miraculously kept alive decades after Beckman stopped making or servicing them. I think this high technology from the pre-computer era still is awe-inspiring and would have deserved a bit more attention, although I can always go back to the 1952 book, which has 10 pages on this.

Among more recent technologies used to study polymers, the upgrading of mass spectrometry from a small molecule method to one that now routinely used on large proteins would have deserved an explanation. For decades, mass spectrometry had been used only for small organic molecules. From my own research experience I know how much effort has gone into teaching polymers how to fly, and to detect the presence or absence of single protons in masses in the megadalton range.

Reaching the recent past towards the end of the book, the author gives quite a few examples of DNA nanotechnology, including the self-assembled DNA patterns and three-dimensional objects, based on papers published in 2006-2008. He repeatedly states that we don't know yet whether DNA constructs will ever be useful in real-world applications. This suggests he missed out on electronic DNA aptamer biosensors which have been demonstrated to outperform commercially available sensors in real life situations and are already quite close to commercialisation.

These moans apart, the book gives an interesting account of the colourful history and present promise of polymers of all shapes and sizes. I believe there should be more books like this, giving the famous intelligent lay reader access to important concepts that have been neglected because they cut across the established structures of scientific disciplines. And there should be more people reading them.

II. Short reviews

This section contains reviews written for different outlets. Many of the early examples were published in Chemistry in Britain, some in its successor Chemistry World, and quite a few, again, in Chemistry and Industry. In the last years of the decade, I also published a few book reviews on my blog, www.proseandpassion.com, which are also included in this section as far as they are related to science.

Sinful thoughts

The genetic inferno: Inside the seven deadly sins
John Medina
Cambridge University Press, 2000

Originally published in Chemistry in Britain, 37, No. 1 (Jan. 2001), 47

Don't panic -- this is not a revelation of another gene-related scare story following in the tracks of GM food. The word "inferno" is meant as a (slightly misdirected) pointer to the fact that Medina takes some inspiration and a lot of metaphors from Dante Alighieri's *Commedia*, the famous *Divine comedy*. The seven deadly sins which Medina uses as a scaffold to organize his subject matter, the biology of human behaviour, are in fact borrowed not from the hellish *Inferno* section of the divine comedy, but from the *Purgatorio*.

Why do we crave chocolate, sex, money, and how do we know we do? What happens at the sensory level and how does it trigger biological signalling chains that end up in our conscious perception? These are the kinds of questions that Medina wants to bring to a non-specialist audience, making the scientific facts accessible by use of metaphors and his own excellent graphics. The trouble with this is that many questions are still unresolved. In the words of the blurb, Medina is "describing the gap that exists between a human behaviour and a human gene". Mind the gap. Of the parts that are known, some are too complex to be brought to a general audience in detail.

What is understood by science and understandable to the general public would have made a slim volume, but publishers want to see around 300 printed pages. Medina's solution to this problem is that he uses Dante not only to mine for metaphors and examples that are sometimes brilliant, but also to stop the gaps and fill the cracks. His oscillations between explanations of hard scientific facts, snippets from Dante, and his own philosopy create a rather complex text structure, which in turn makes it necessary for him to add extensive signposting. A significant length of text is actually spent on crossreferencing and explaining his route.

If you want to enjoy this book, you should bring a solid interest in both human biology and mediaeval literature, and an above average attention span.

Back to basics

Basic biotechnology
Colin Ratledge and Bjørn Kristiansen (eds.)
Cambridge University Press

Originally published in Chemistry in Britain, 37, No. 8 (Aug. 2001), 44.

Biotechnology is one of those words that mean different things to different people. When you hear it out of context, do you think of (tick as appropriate):
* beer and wine
* antibiotics
* GM food
* quorn
* drug discovery
* Dolly the sheep
* others - please specify: .... ?
The diversity of this field, together with the different perceptions of it ranging from enthusiasm to strong opposition against some of its parts, makes it difficult to write a textbook about it. Ratledge and Kristiansen have assembled more than 30 experts from various branches and different walks of life to help them cover most aspects of this broad subject matter, divided into two main parts: Fundamentals and principles (ca. 250 pages) and Practical applications (ca. 300 pages).

They start out with the public perception and image problems of biotechnology. I would have preferred to read a clear definition of the field, perhaps aided by a brief historical overview of its predecessors, at the beginning. In order to understand why there are misgivings and misconceptions about a technology, one needs to know first of all what it's about. The current chapter 1 (which might have been better placed at the end of the book) fails to build any bridges between biotechnology and its opponents, showing very little understanding for the concerns of the latter. Some issues are also meddled unhelpfully. Citing insulin as a major success of genetic manipulation is misleading, as most insulin-dependent diabetics can be treated with the hormone purified from pig pancreas, and millions of lives have been saved by this old-fashioned pre-GM technology.

Part I then goes on to outline the essentials of biochemistry and cell biology as far as they are needed for the understanding of technical applications. These chapters are concise and well thought through. Part II kicks off with a cheerfully written introduction into the biotech business, complete with helpful hints for scientists who may want to turn their "great idea" into a startup company.

The remaining chapters are mostly product oriented, dealing with subjects such as amino acids (including sodium glutamate of Chinese takeaway fame), antibiotics, or enzymes. In the the otherwise very comprehensive chapter on antibodies, I missed a reference to the promising new field of using derivatives of the heavy-chain antibodies found in *Camelidae* (camels, llamas, etc.).

All in all the book is well presented, with clear and instructive diagrams and useful glossaries for those subject matters that come with their own specific nomenclature. It will certainly be a good resource for anybody who wants to study biotechnology or move into the field from another discipline.

Entertaining tales

Stories of the invisible: A guided tour of molecules
Philip Ball
Oxford University Press 2001

Originally published in Chemistry in Britain, 37, No. 12 (Dec. 2001), 41.

The world of the "Mollycules", as they are called in a charming extract from Flann O'Brien's *The Dalkey archive* which opens Ball's book, is by its very nature invisible, and thus inaccessible to many. Philip Ball is offering his readers a guided tour of the molecular world, to show them things they can't see, to make them grasp the things they can't touch. His tour has seven stages focusing on synthesis, biomolecules, materials, energy, motors, communication, and information.

On the biological fringe of his subject, some inaccuracies have crept in. I believe the sections about the invention of oxygen producing photosynthesis should have credited the cyanobacteria rather than algae (p. 102 and 105). The directionality of tubulin growth has been muddled (p.122), and I am not sure I believe the claim on p. 118 that bacterial life wouldn't be possible without dedicated motor proteins (not counting DNA polymerase and similar moving enzymes).

Each chapter has some light entertainment and cross-cultural references thrown in, like O'Brien's "Mollycule Theory", or Primo Levi's well-known memoir *The periodic table*. The bulk of the material, however, is a highly condensed description of current knowledge in the molecular sciences. "Chemistry without the boring bits" one might have called it. But *Stories of the invisible*? I haven't spotted enough storytelling in this book to justify this title. There are wonderful stories hidden in the history of chemistry up to this day, stories of discovery, serendipity, impossible quests, but here they disappear behind the molecules which are the main focus of the book.

This leads us to the dreaded question of the target audience. Like many of the few chemists writing for a general audience (this reviewer included), Ball has a clear agenda of what he thinks the general public should know. The question remains: does the general public want to know what chemists want them to know? And if not, how can one smuggle it into their brains anyway? Telling real stories is a route that works but is rarely used in chemistry. But as long as we chemists insist on putting the science up front and the stories in the background, our audience will probably remain limited to those people already scientifically educated.

Happily ever after

Girls, Genes and Gamow
James D Watson
Oxford University Press 2001

Originally published in Chemistry in Britain, 38, No. 3 (Mar. 2002), 76.

Once upon a time, an irreverend 24-year-old American biology postdoc and a mildly eccentric British physicist worked together at the Cavendish laboratory at Cambridge. They tinkered with models to find a structure that fitted the X-ray diffraction patterns that others had measured of DNA, and they discovered the double helix. They became very famous and lived happily ever after. This is the story James Watson has told the world in his famous memoir "The double helix", first published in 1968 and translated into numerous languages. Although one may argue about the way he treated some of the people involved, including the late Rosalind Franklin, it has deservedly become a classic of science writing.

Now the author presents a sequel, to give us some details of the "happily ever after", and to back up his claim that, thanks to the double helix, he has led an extremely interesting life in science. The only trouble is: the data he presents here don't support this claim. What he did next was to spend a year at Caltech, making theories about how DNA sequences lead to proteins, none of which worked out (while Crick proposed the adaptor hypothesis, which did: adaptors are now known as tRNA). Genes? What genes? Not much more luck at the girls front. Given that he characterizes the women in his life mainly by their hair colour, and expresses surprise that Christa Mayr, daughter of the acclaimed biologist Ernst Mayr, could keep up her end in serious conversation, readers may feel like congratulating those that got away.

And finally "Geo" Gamow: yes, he must have been an interesting character, but he doesn't really come alive on these pages. The same holds for Richard Feynman whom Watson met regularly. (Now that's at least one person who had a more interesting life in science.) Still, writers can get away with flat characterizations of their cast as long as they have a plot. What makes this book fail where the double helix succeeded is the nearly total absence of a structured plot. Its content can be summarized in 15 words: famous but frustrated scientist looks for girlfriend and faculty job, finds both in the end.

Transforming bioenergetics

Wandering in the gardens of the mind: Peter Mitchell and the making of Glynn
John Prebble, Bruce Weber

Oxford University Press 2003

Originally published in Chemistry in Britain, 39, No. 7 (July 2003), 53.

This biography comes with a very useful feature - an executive summary at the beginning of the book. The seven pages of the "Prologue - Who was Peter Mitchell?" are short enough to read in a bookshop cafe, and they tell you everything you really need to know about the 1978 Nobel laureate for chemistry. Which are, in a nutshell, two things. He transformed bioenergetics by coming up with the chemiosmotic theory, which essentially states that cells can store energy in the form of ion gradients across a membrane, which can be converted into biochemical energy by the ATP synthase. The second take-home lesson about Mitchell is that he was an artistically minded, mildly eccentric scientist from an extremely wealthy family, which enabled him to leave the academic system when he had enough of it and to buy a manor house to found his own research institute, the Glynn Research Laboratories featured in the title of the book.

Should the prologue tickle your fancy sufficiently to take the book home, you will find a thoroughly researched and documented biography of somebody whose mind was certainly wandering in a number of directions. While the abundant details of Mitchell's life and research may concern only the historians of science, I think that chemists and the wider public can learn from this biography that even wildly meandering paths through science can lead to success and a Nobel prize. Mitchell had his first PhD thesis rejected, didn't get on with his head of department, and had an inclination towards somewhat speculative theory, which was only balanced out by the experimental skills of his longstanding collaborator Jennifer Moyle. Luckily, he turned his unusual way of looking at things at a topic which at that time needed a paradigm shift.

Read this biography as a proof - if proof were needed - that all scientists don't think the same way, and that scientific progress can benefit enormously from those who stray off the garden path.

Small is beautiful

Lab-on-a-chip: Miniaturized systems for (bio)chemical analysis and synthesis
R.E. Oosterbroek and A. van den Berg (eds.)
Elsevier 2003

Originally published in Chemistry World, 1, No. 5 (May 2004), 59.

Liquid handling on a micrometre or even nanometre scale is a relatively new concept that was first proposed and tried out just over a decade ago. In the last few years, it has made spectacular progress (see Chemistry World January 2004, p. 26), such that a number of applications have already been realised, ranging from the analytical (mass spectrometry) through to the synthetical (micro-PCR).

The volume compiled by Oosterbroek and van den Berg provides a first summary of the state of the art in this emerging technology. It is organised into four parts: the first deals with the different materials that can be used, i.e. hydrogels, polymers, silicon and glass, and then moves on to more application-oriented concerns, such as surface chemistry and production technologies. The second part deals with different methods of handling liquids in the microscale, while the third focusses on the special cases of bead-based applications and cell-counting systems. The final and longest section then leads us into the promised land of real-world applications, dedicating chapters to drug delivery chips, capillary electrophoresis, and mass spectrometry among other fields.

Given that the book is essentially a scientific monograph consisting of 18 review articles by experts from different institutions, it is surprisingly readable. Strict editing, the sensible organisation of each chapter with sections including a gentle introduction and a summarising conclusion, and generous illustration with instructive diagrams and black-and-white photos allow the reader to enjoy this work more thoroughly than most of its kind. The only thing missing is a general introductory chapter to the whole volume.

In fact, one could almost recommend it to the more ambitious members of the lay readership, although the exorbitant price will probably limit the audience to those who are really keen to find out about all the wonderful things you can do on the small space of a chip.

Analysing with proteins

Protein microarray technology
Dev Kambhampati (ed.)
Wiley-VCH, 2004

Originally published in Chemistry World, 1, No. 5 (May 2004), 59.

The first commercial DNA microarray went on sale ten years ago. Since then, the relatively straightforward production and handling of nucleic acid arrays, combined with its evident usefulness in applications like high throughput screening has made this technology a major success story. In the age of proteomics, it would of course be highly desirable to have the equivalent microarrays equipped with proteins to test for protein-protein interactions on a proteome-wide scale. Unfortunately, the fact that proteins are both chemically more diverse and more sensitive to environmental conditions than nucleic acids means that they are much more difficult to stick into a microarray. With the notable exception of antibodies, most kinds of proteins still present a major challenge to any array-style application.

Accordingly, this monograph of 10 chapters from different laboratories is a report on work in progress more than an overview of achievements. It deals with a variety of techniques that are either leading towards or vaguely located in the neighbourhood of protein arrays. Three chapters, including the longest one (covering 50 pages) deal with surface plasmon resonance detectors, two more with readout methods including cantilever sensors and image analysis. The remaining five chapters include a general (and helpful) introduction, two about the surface chemistry required for arrays, and two about protein microarrays as such.

The heterogeneous nature of the collection may reflect the state of a field still struggling with basic technical problems. However, it makes for a book that would be hard to read from cover to cover. It will probably be most useful for those who want to get into the field and need a thorough briefing on the available methods and current challenges.

Missing out on interdisciplinarity

Nanoparticles: From theory to application
Günter Schmid (ed.)
Wiley-VCH, 2003

Originally published in Chemistry & Industry, No. 11 (May 2004), 25.

The exploration and utilization of the nanoworld is by its very nature an interdisciplinary endeavour. Biological systems can provide inspiration on many different levels to those who aim at ordering and using matter on the nanometre scale. Nano engineers can, for instance, borrow biological molecules from the cell, they can engineer them for their own specific needs, and create new interfaces with electronics or solid state physics. At yet another interface, chemists have used the biological principles of self-assembly and weak interactions to create supramolecular chemistry.

It is therefore to be deplored as a missed opportunity if -- as has happened in some places -- new “nano” institutes (e.g. of nanochemistry, nanomaterials or nanoelectronics) fail to create infrastructures for broad interdisciplinary approaches and instead just turn out to be sexed-up departments of chemistry, materials or electronics. Similarly, this book is a missed opportunity to create dialogue between the disciplines that have connections to the concept of a nanoparticle. It might have placed the semiconductor and metal nanoparticles that are clearly the key interest of the authors in the context of chemical and biological structures, thereby bridging the gaps between the disciplines.

Instead, this appears to be a book written by solid state physicists/chemists exclusively for their own community. The only concession to other disciplines is the sixth chapter which discusses biomaterial-nanoparticle hybrid systems. It creates the impression that biomaterials are interesting for the sole reason that one can patch them onto nanoparticles, and the whole chapter could be described as a patch to cover up the lack of interdisciplinarity in this book. Advanced students of physics might find the volume helpful, but even they deserve to be told that there is more to life than solid state physics.

“Where is Waldo?” goes nano

Bionanotechnology: Lessons from nature
David S. Goodsell
Wiley-Liss, 2004

Originally published in Chemistry World 1, No. 8 (Aug. 2004), 51.

Textbooks tend to show the cell as a cuboid with rounded edges containing only a few neatly arranged protein complexes. While graphic oversimplification is often necessary for didactic purposes, one should never forget that real life cells are anything but neat and simple.

For anybody who wants to know what life at the nanometre scale really looks like and doesn't have an electron microscope at hand, David Goodsell's illustrations are the best window into the nanoworld. In 1993 he gave us "The machinery of life", including fantastic drawings that made the emerging concept of molecular crowding in the cell a palpable reality. The cover of the book alone -- a cross section of a bacterium, magnified 200,000 fold and jam packed with ribosomes and other machinery -- makes the viewer wonder how anything inside a cell can move at all. Finding the right molecule to interact with is in fact like finding Waldo in the overcrowded pages of the popular children's books.

With "Bionanotechnology", structural biologist Goodsell now builds a bridge to applied science and presents the molecular machines of the cell not only with the much improved structural insight available a decade after the first book, but also with a view towards making the understanding of their mechanisms useful for our own nascent attempts at controlling matter on the nanometre scale.

At first glance, the resulting volume looks like a compact textbook, particularly due to the chapter subheadings that summarize the contents of the following paragraphs in a short sentence (e.g.: "The Hydrophobic Effect Stabilizes Biomolecules in Water"). But with its stunning two-colour drawings and a highly readable text, one should hope that this book will not only educate students but also reach a wider audience.

In praise of blue skies

Pioneering research - a risk worth taking
Donald W Braben
Wiley Interscience 2004

Originally published in Chemistry World 1, No. 10 (Oct. 2004) 75.

Is science becoming predictable and boring? Does it no longer produce pioneers like Albert Einstein or Barbara McClintock who think outside of the many boxes provided for them and eventually change the way we view the world?

Donald Braben, a physicist with a colourful career spanning research, administration and industry, certainly thinks so. In this book, he argues vehemently that the growing obsession with target-oriented, focused, plannable work (which of course is necessary as well) is pushing the truly creative and exploratory science over the edge. The biochemist Howard Schachman once described this phenomenon in a diagram which I like to call the Schachman plot in his honour: he plotted funding probability against originality of grant proposals. The curve starts with a modest funding probability at zero originality, rises to a maximum at an average originality, and then drops sharply to zero where the proposals become 'too creative'.

Braben's concern is exclusively with those 'blue sky' researchers who fall off the sharp end of the Schachman plot. Global progress is at risk he argues, if we only pursue the kind of research that can be predicted and fitted into five-year plans. He argues that concepts like competition, efficiency, peer review and market orientation should not apply to pioneering research. Creative scientists should be unique, peerless individuals, like creative artists, not athletes in a race. A small number of them should be given total freedom to follow their ideas. In the concluding chapter, Braben writes the history of BP's Venture Research Unit, which under his leadership strived to do just that during the 1980s, but ceased to exist in the 'focused 1990s'.

While many of Braben's views are heretical in today's Church of Efficiency, I for one agree with much of his assessment and recommend this book to anyone involved in or worried about science.

Catalysis uncovered

Concepts of modern catalysis and kinetics

I. Chorkendorff, J.W. Niemantsverdriet

WILEY-VCH, 2003

Originally published in Chemistry & Industry No. 24 (Dec. 2004), 24.

The importance of catalysis cannot be overstated. Take biologically available nitrogen, for example. It has been estimated that roughly half of the nitrogen feeding the world population comes out of the Haber-Bosch process. Without this catalytic process, the foundation of which Fritz Haber worked out a century ago, starvation would have capped population growth, and today there would be several billions less of us. The other half of the nitrogen comes from natural sources. But these, too, require catalysis to convert molecular nitrogen to usable nitrogen. Nature's Haber-Bosch process is an enzyme called nitrogenase. Without catalysis, there would be no life. While chemical reactions would still take place, their sum total would amount to geology, not biology.

And yet, many chemists seem to take catalysts for granted, like lab instruments that you take from the shelf when you need them. At the various universities where I studied, at least, I never heard of a course unit on catalysis. In my studies, catalysts cropped up here and there, when they were needed to speed up inorganic or organic reactions, such that they could be completed within the time scale of a laboratory course. Physical chemists would explain to us that catalysts conveniently lower the activation barrier of a reaction. And that was it -- catalysis had fallen between the cracks of the traditional separation of disciplines.

It appears that today's students are better off, at least if they study at the technical universities of Eindhoven (Netherlands) or Lyngby (Denmark), where Niemantsverdriet and Chorkendorff teach. The present textbook on (heterogeneous) catalysis and kinetics is based on the courses they have taught over many years.

It starts off with a highly readable general introduction to catalysis, followed by chapters on kinetics and reaction rates which cover about the same ground that would also be included in general textbooks of physical chemistry. Then there are chapters on how to characterize (solid) catalysts and what one should know about various types of these. Two chapters deal with the reactions on surfaces, and another three tackle application-oriented aspects of the field, namely environmental catalysis, oil refining, and hydrogen-based processes including the ammonia synthesis.

All in all, the book is well-written and richly illustrated with instructive (black and white) diagrams. Students, especially those at Eindhoven and Lyngby, will have to read it from cover to cover, of course. More mature readers like myself have the privilege to skip the dry bits on kinetics (been there, done that, got the T-shirt) and focus on the catalysis parts for the sheer excitement of the facts. If you always wondered how the three-way converter on your car works, or what happens to the crude oil in a refinery, the last chapters of this book hold all the answers. Even though homogeneous and enzymatic catalysis are underrepresented, the book does catalysis justice as a central concept of chemistry.

Guide for the inexperienced

Guide to analysis of DNA microarray data (second edition)
Steen Knudsen
Wiley-Liss 2004

Originally published in Chemistry & Industry No. 2 (17.1.2005), 20.

The two fundamental ideas which which eventually led to the development of nucleic acid and protein arrays must both be celebrating their 30th birthday round about now, as they were both published in 1975. One was the step from one-dimensional to two-dimensional electrophoresis, pioneered by cell biologist Patrick O'Farrell, which allowed researchers for the first time to visualise the entire protein production of a given cell spread out over a surface as a multitude of black spots, separated according to the protein size and charge. From this work and the rapidly spreading application of 2D electrophoresis, it became clear that the analysis of the gene expression of a cell needs two dimensions and quantitative analysis of the signals spread out on the gel or array. The other 1975 breakthrough was the Southern blot, named after Ed Southern, who transferred DNA from a separation gel onto a membrane, where the nucleic acid molecules stayed immobilised in their separated spots and could be probed with complementary, labeled RNA strands.

In the 1990s, the two-dimensional visualisation of gene expression and the hybridisation of immobilised target nucleic acids with mobile, labeled probe oligonucleotides merged into the development of the array, of which there are now a wide variety on offer for anybody who wants to analyse gene expression or screen for specific sequences. As is to be expected in a rapidly growing field, there are many people who will want to use arrays but have no experience with them. Hence the need for guide books such as Knudsen's.

This guide is mainly about the analysis of the data obtained from DNA arrays, but it starts with a useful 18-page introduction into the types of array technologies available today and their respective strengths and weaknesses. This part of the book could be useful for anybody who wants to catch up with the marvels of modern lab technology, even if they don't plan on using it. In the following chapters, the book gets rapidly more technical, outlining fundamental considerations regarding the localisation, identification, and quantitative analysis of spots obtained from imaging an array, the usefulness (or otherwise) of background subtraction, statistical methods to assess the significance of any intensity changes observed, and so on.

Each of the 12 chapters on data analysis is concise and fairly readable considering the mathematical nature of the subject. Each comes with several pages of "further reading" grouped by specific topics. At the end of the book, the same references are listed again, in alphabetic order, all 207 of them. Which means that, without the references, graphs, tables, and bullet-pointed lists of computer programs or such like, there remain much less than 100 pages of flow text to read. This explains why the booklet appears quite accessible to the non-expert, but it also suggests that the more experienced users of array technologies may find it too slim. Thus, it is probably more of an introduction than a guide, most suitable for newcomers to the field of DNA arrays.

The mature ribosome

Protein Synthesis and Ribosome Structure
K.H.Nierhaus, D.N.Wilson (eds.)
Wiley-VCH 2004

Originally published in Chemistry World 2, No. 2 (Feb. 2005), 56.

The machinery that translates the genetic information into amino acid sequences has posed a major challenge for molecular biology. Since the 1960s it has been clear that the ribosome, a large complex of RNA and more than 50 different proteins, is the key component whose structure and molecular function had to be cracked. Traditionally, the ribosome community met every three years and produced a hefty proceedings volume containing more questions than answers.

After the gold rush of crystal structures published from 2000 onwards, ribosome research has matured into a field where the fundamental questions can be answered in molecular detail. Knud Nierhaus, a pioneer of the research into ribosome function and assembly, and his younger colleague Daniel Wilson (both based at the Berlin Max Planck institute) have edited the first ribosome book that reflects this new situation and gives the answers to the questions thrown up in the earlier books.

The book opens with an overview of ribosome (pre)history by ribosomologist-turned-historian H.-J. Rheinberger, followed by an account of the current structural knowledge by G. Blaha. The next 10 chapters are dominated by the editors' research interest, namely function and assembly of the ribosomal complexes, and between them they have co-authored most of this main part of the book. J.-H. Alix rounds it off with a chapter on molecular chaperones, which take care of the nascent polypeptide chain emerging from the ribosome.

With its thousands of references, this is clearly a scientific monograph, but given the new-found maturity of the field, it should also be a useful resource for everybody with a more marginal interest in protein synthesis, condensing five decades of research into a manageable and indeed readable volume.

Nano A-Z

What is what in the nanoworld: A handbook on nanoscience and nanotechnology
Victor E. Borisenko, Stefano Ossicini
Wiley-VCH

Originally published in Chemistry World 2, No. 2, 58.

Nanotechnology has finally arrived in the mainstream of science. One of the symptoms of its growing importance in real life (as opposed to science fiction) is the deluge of nano-titled academic books published last year. As some authors clearly take advantage of the nano bandwagon to promote the parochial views of their particular subdiscipline without much care for the emerging new qualities of nanotechnology, these books have to be approached with caution.

"What is What" is essentially a dictionary with definitions arranged alphabetically. Entries vary in length between one-liners (interface - a shared boundary between two materials) and short articles of up to five pages length (e.g. carbon nanotube). Only scientific concepts are listed, no scientists. However, many of the longer articles finish with the name of the inventor / discoverer of a concept, and any publications / recognition related to it.

The choice of "nanoworld" keywords listed in the alphabet is dominated by the authors' background in physics, leaving biology and chemistry underrepresented. While physical concepts of unclear relevance to nanotechnology, including "neutron" and "Schrödinger's cat" find space in this book, biological entities like microtubuli or myosin, and chemical concepts like supramolecules and dendrimers are missing. The style of the longer articles (e.g. carbon nanotube) confirms that this is a book written by physicists for physicists, without any attempt at interdisciplinary integration.

Thus, it appears that the present volume is a short dictionary of 20th century physics, updated with some key concepts of nanotechnology. By calling it a "handbook" in the subtitle, the publishers unwisely invite comparison with the "Springer Handbook of Nanotechnology" (Springer Verlag 2004), which offers much better value for money.

Soft landing

Soft Machines: Nanotechnology and life
Richard A. L. Jones
Oxford University Press 2004

Originally published in Chemistry World 2, No. 4 (Apr. 2005), 56.

There are two different approaches to overcoming the gap between science and the general public. Many scientists, myself included, have been trying to popularise what they think the people should know about science, building their bridge from the science side of the canyon. Fewer, but much more successful in terms of bestsellers lists are the examples of non-scientists like Bill Bryson or scientists converted to populism, who build the bridge from the other shore, starting from what non-scientists actually want to read.

Soft machines is a beautiful example of the former school of thinking, a nicely written account of what a physicist thinks the public should know about nanotechnology. Richard Jones, a polymer researcher and professor of physics at the University of Sheffield, leads us into the nanoworld from the physics entry, starting with imaging and fabrication methods, then explaining what makes nanoscale mechanics so different from the world we know. He then comes to the core of the book, two chapters on the "soft machines" of the title, and another two on computing and electronics on the nanoscale. He rounds off the book with a very short chapter on future opportunities and risks.

The book serves up a fair amount of real science, made palatable with original metaphors and a light sprinkling of anecdote. Thus, it is ideally suited for us chemists, as we are close enough to physics to understand the odd equation that Jones throws in, and distant enough to benefit from this change of perspective. But for the non-scientists whose scientific education is on the level of "A short history of nearly everything", this book is too hard. I wish the general public would make an effort to read such books, but my royalty statements tell me loud and clear that they don't.

Nanotech for students

Nanoscale science and technology
R. Kelsall, I. Hamley, M. Geoghegan, eds.
Wiley

Originally published in Chemistry World 2, No. 7 (July 2005), 57.

The science of small things is turning into big business. It appears to be lucrative not only for the makers of nanostructures that end up in optical or electronic devices, but also for the publishers who produce an ever increasing stream of books with the four letters "nano" on the cover. After the early vision books with their prophecies, we have had the popular science books, then the handbooks and academic monographs, and now there is the first (at least on my shelf) student textbook on nanotech.

Like most textbooks, it is based on the editors' own teaching experience, namely on the pioneering "Nanoscale science and technology" masters course they have been running as a joint project of the Universities of Leeds and Sheffield since 2001. Like the course, the textbook is targeted at postgraduates from the physical sciences.

The nine chapters contributed by eight different authors are arranged vaguely from the simple to the complex. The book starts with fundamental things from atoms upwards, introduces both production and analytical techniques, and the basics of semiconductors. In the last third we find self-assembly, molecular films, and finally the most complex nanosystem known to mankind, the living cell, with some of its machinery.

Overall the book reads well, abounds with instructive diagrams, and appears to be pitched at the right level for a postgraduate audience. As it is clearly rooted in the physical sciences (the "bio"-chapter is the shortest of the book and only offers a few eclectic examples of bio/nano cross-fertilisation), it leaves an opening for a more biology-based text to help any biologists or even medical students who choose a nano subject in their postgraduate studies. Once that has been done, we only need the children's guide to the nanoworld in order to complete the nano bookshelf.

Cool science

Cryogenic engineering (second ed.)

Thomas M. Flynn

Marcel Dekker

Originally published in Chemistry & Industry No. 18 (19.9.2005), 27.

What do the Saturn rockets of the Apollo program, the MRI scanner at a hospital, and your fridge have in common? They have all been built using cryogenic engineering, or the art of making things very cold. Which is the technology you need to get liquid gases, superconducting magnets, and common household refrigerators.

So if cryogenic engineering is clearly useful and important for such a broad range of applications, why aren't there stacks of books about it, and why isn't it taught as a degree course at universities? I guess the answer may have to do with the fact that the applications of cryogenic engineering are so different, some of them old and well established, some of them rather exotic. The cooling technologies used in fridges haven't changed in principle for a century or so (only the liquids used have changed for various reasons). And the technologies to build rockets with liquid gas as fuel or superconducting magnets are only relevant to a small number of people.

It's for these people "who work for a living" (as the author defines them in the the foreword, presumably to set them apart from mere pen-pushers like me) that Thomas Flynn has compiled this book from his decades of working experience in the field. He sees it "as the cryogenics book I would like to have on my desk," i.e. a useful reference rather than a teaching aid or a reading experience.

Nevertheless, he conjures up a fairly readable introduction to the history of cryogenics, rich with the cute factoids that would make this story a rewarding subject for a popular science writer. We learn, for instance, that natural ice was at one point the second most important export of the United States (after cotton), and the true reason why beer is consumed very cold in Germany and rather warm in Britain. The second chapter is a short introduction into thermodynamics, with special emphasis on cooling matters.

Then the handbook character of the book takes over, and it's facts, facts, and even more facts. The third chapter, by far the bulkiest with 180 pages, contains everything there is to know about cryogenic fluids, summarised in 66 tables, 142 figures, and just a little bit of text in between. The following two chapters deal with the relevant properties of solids, while each of the remaining six chapters addresses a specific aspect of cryo-work, e.g. how to liquefy gases, insulation, refrigeration techniqes, etc.

While I am not in a position to check any of the factual content, I certainly would want to have this book on my desk if I ever decided to build a rocket, a big magnet, or indeed a fridge.

Chemistry in all its beauty

Elegant solutions: Ten beautiful experiments in chemistry
Philip Ball
Royal Society of Chemistry 2005

Originally published in Chemistry World 2, No. 9 (Sep. 2005), 73.

When Britain's TV audience voted for the 100 Greatest Britons, chemists found themselves in the margins of society, clearly less valued than politicians, artists,. and scientists from other disciplines. Similar voting exercises held in Germany, Finland, Canada, the Netherlands, and the US make similarly depressing reading for the chemist.

Flying the flag for chemistry, both the American Chemical Society and the RSC have independently set out to honour the greatest experiments in chemistry. The latter has commissioned science writer Philip Ball to compile his personal favourites into a book.

His choices, partially overlapping with those of the ACS members, cover a broad range from atoms and their decay (Rutherford, Curie) through to biological chemistry (van Helmont's willow tree, the Miller-Urey experiment). I have described four of the ten experiments in my own books, at least in part because of their sheer beauty, so my own selection would overlap with Ball's as well. Pasteur's crystal sorting experiment, which for the first time established a relation between a macroscopic property with a molecular one, is an eternal favourite that would probably appear on anybody's list. If I had to add an eleventh experiment to Ball's selection, it would probably be Ghadiri's self-assembling peptide nanotubes.

The book makes good holiday reading (as I tested experimentally), and is fully compatible with trains, planes and hot weather. Will it change the outcome of the next "Greatest Britons" poll in favour of the chemists? I doubt it, but for a silver lining look at the result of the French poll: Marie Curie and Louis Pasteur, both featured in this book, came in fourth and second, respectively. So maybe there still is hope for beautiful chemistry to be recognised alongside the other aspects of our culture.

Biochemistry rediscovered

Chemical biology -- A practical course
Herbert Waldmann, Petra Janning
Wiley-VCH 2004

Originally published in Chemistry & Industry No. 24 (19.12. 2005), 24.

Once upon a time, a long time ago, biochemists didn't know about molecular biology. When they set out to investigate a protein from potatoes, they actually started by homogenising a sack of potatoes, and went on to purify the protein from that, rather than expressing the gene from some nondescript bacteria. Similarly, if they wanted to study eye lens proteins, they got a load of calf's eyes from the local slaughterhouse ... well, I'll spare you the rest.

Back then, biochemists were also seen practicing actual chemistry, creating new inhibitors for their enzymes and studying their binding kinetics. Those times ended rather quickly when the spread of molecular biology technology steamrollered most of biochemistry into a very flat, gel-shaped state.

Now there is a new inter-discipline emerging, known as chemical biology. What exactly is it about, and how does it differ from biochemistry? I admit that I was puzzled at first and agreed to review this book because I hoped for an answer to these questions. I figured out from the introduction of the book that chemical biology is essentially a revival of biochemistry as it was practiced before molecular biology took over the labs.

Thus, one of the 12 experiments described in the book involves the purification of phosphorylase from 2.5 kg of actual peeled potatoes. It forms part of a very diverse mix of techniques, ancient and modern. Apparently, everything that can serve to do biochemistry but isn't molecular biology is deemed part of chemical biology. Peptide synthesis, mass spectrometry, fluorescence techniques, oligonucleotide synthesis, peptide nucleic acids, in silico ligand design, lipidation of proteins, combinatorial chemistry, are just some of the topics covered.

Now the crunch question is the following: Is chemical biology "a new science" as the authors claim in the introduction? Will it establish itself as a new discipline, with dedicated institutes, professorships, textbooks, practical courses and all that? Even though I'm very much in favour of giving this kind of research more attention, I'm not sure whether it qualifies as a discipline. It is a set of methods all right, but if you consider the results obtained with these methods, they naturally fall into the remit of biochemistry. While there may be chemical biology methods, there will be no body of knowledge that could be summed up in a textbook called "chemical biology" and would differ in any way from the knowledge we call biochemistry.

As a set of methods, it still deserves university courses, and possibly a few books to go with them. This one admits to being a tentative approach, based on the teaching and research of the editors and their co-authors. It is certainly not going to be the last word in this field, but if it helps to inspire other academics to develop their own practical courses at their universities, possibly with different emphasis depending on the range of technologies and experience available, it will have been worthwhile.

Haber revisited

Between genius and genocide: The tragedy of Fritz Haber, father of chemical warfare
Daniel Charles
Jonathan Cape

Originally published in Chemistry World 2, No. 12 (Dec. 2005), 59.

Nitrogen from the air can enter the chemical cycles of the living world via two processes that globally turn over similar amounts. One is bacterial nitrogen fixation; the other is the synthesis of ammonia from the elements under high pressure, invented by Fritz Haber.

Apart from the process that literally feeds today's world, Haber is also known for his enthusiastic services to the Reich of Kaiser Wilhelm during WW I, particularly in the development of chemical warfare. With the arrival of the Third Reich, however, Haber found that his Jewish origins outweighed his achievements and services. Soon he found himself exiled, and he died before he could find a new home.

The moral complexity and tragic conclusion of his life make Haber a tricky subject. His former assistant Johannes Jaenicke spent decades collecting materials for a biography, but never got it written down. His collection is the archive from which biographers feed, including Dietrich Stoltzenberg (whose epic effort recently appeared in a shortened translation, reviewed in Chemistry World, Feb. 2005, p. 55) and now Daniel Charles.

Charles's advantage is that he sees Haber with the fresh eyes of an outsider, who admits that he once visited the Haber institute without knowing who it was named after. Since then, he has certainly done his homework at the Jaenicke archive, and manages to tell the story in a compelling and fascinating, yet compact and accessible form. His strength is the witty summary ("Haber didn't immediately volunteer for this epic quest. He had to be goaded into it with offers of money and insults to his pride." -- p.83) that often introduces a new section of his story. The only blemish is the title of the book, which demonises the creator of one of the most important inventions of the 20th century.

PS: On the subject of Haber, I have also written a long essay review (see "For better, for worse, above) and an opinion piece which I'm attaching here as bonus material:

Elusive treasures

Originally published in Chemistry in Britain.

If it comes to playing "six degrees of separation" with chemists, I happen to be reasonably well positioned. For example, the monumentally tragic figure of Nobel laureate Fritz Haber (1868-1934) is only three steps away, as my former PhD supervisor is a son of Haber's assistant and would-be biographer Johannes Jaenicke. I cherish this connection as it throws a somewhat unusual light on the great man.

The well-known parts of Haber's biography are all collossal in their impact on history and people's lives. Being a very patriotic German citizen during WW I, Haber helped to develop chemical weapons and oversaw their use at the front. Although aware of the horrendous suffering they caused, he reckoned they would bring a quick end to the war and thus reduce overall loss of lives. His young wife, Clara Immerwahr, one of the first women to gain a doctorate in chemistry, disagreed and committed suicide after failing to stop him. On the other side of the balance, one can estimate that roughly half the nitrogen feeding the world population today comes out of the Haber-Bosch process he invented. Bringing a complex biography to a tragic end, he had to flee Nazi Germany and died in Switzerland on the way to Rehovot, where Chaim Weizmann had offered him a position.

Haber's former assistant Johannes Jaenicke and his wife spent decades collecting materials for a biography, which due to the sheer size of the task and his failing eyesight in old age, he never finished (the recent biography by Dietrich Stoltzenberg is largely based on Jaenicke's material). One research project that Jaenicke was involved in himself took place between the wars, and offers a little light relief in the Faustian life of Haber. In the spring of 1920, Haber surprised Jaenicke and other coworkers with the announcement that he wanted to investigate the possibility of extracting the gold content of sea water, which was estimated to be 5-10 ppb. In that case, he reckoned, extraction of significant amounts of gold should be economically feasible and might help Germany pay back the crushing debts resulting from WW I and the Versailles treaty.

Haber set up a group led by Johannes Jaenicke to develop this project further. During five years, up to 12 research staff and PhD students worked on the gold project under strict secrecy. While industrial sponsors including Degussa were informed, the project was kept secret from the Allied authorities. Initially, the researchers worked on improving the analytical and separation techniques. While there was only a limited number of sea water samples available, their analyses seemed to confirm that the gold content was in the ppb range. For the separation, they tested different approaches and eventually settled for binding the gold to colloidal sulphur and filter with sand also charged with sulphur.

By the summer of 1923, the experiments were ready to be transferred onto an actual ocean-going ship. Still under strict secrecy, a group of researchers including Haber himself boarded the passenger vessel *Hansa* bound for New York, officially registered as crew members. Legend has it that Haber was greatly amused to be called up as a "supernumerary accountant". While the chemists worked behind closed doors, rumours among the passengers ran wild, and upon arrival in New York, a newspaper reported: "German scientists see way to drive ships by using mysterious force." A second expedition in the autumn of the same year took Haber's team to Argentina.

The samples analysed during these two trips gave very different results, such that the researchers had to go back to the lab and improve their analytical methods even further, excluding both gold gains (from common chemicals, dust, or jewellery) and losses (e.g. from adsorption to equipment). Thousands of bottles were filled with sea water samples from all corners of the globe, stacked in purpose-built wooden chests, and shipped to Berlin for analysis. However, the disappointing result was that the actual gold content in most samples was two orders of magnitude lower than the early analyses had suggested, which ruled out any hope of economic retrieval. In 1927, Haber published the results and buried the project.

While it failed to reach the goal, however, the project did improve analytical methods, and also helped to secure the funding of Haber's research in difficult times. Even more ironically, if the project had delivered large quantities of gold, the devaluation of the precious metal might have backfired. As the Spanish found out after plundering the gold of Latin America's native population, "finding" tons of gold doesn't automatically make you rich in the long term. Nitrogen from the air was less glamourous, but ultimately more valuable.

Small wonders

Nanofabrication towards biomedical applications
C.S.S.R. Kumar, J. Hormes, C. Leuschner, eds.
Wiley-VCH 2005

Originally published in Chemistry & Industry No. 1 (2.1.2006), 20.

Nanotechnology has evolved into a scientific discipline with real world applications. The most immediate applications are in microelectronics, which naturally shifts to the smaller realms of nanoelectronics, assisted by progress in optical data processing at ever smaller scales. In this area, nanotechnology research has already begun to have an impact on our lives. However, most of us may not have noticed this technological revolution, as we have come to take the exponential improvements of computer power for granted.

The second area where nanotechnology has inspired high hopes for immediate applications is medicine. Unlike computing, medicine has not experienced a consistent trend to make everything much smaller and much faster. However, considering that the ultimate target of most medical efforts should be the living cell, there is a case for medical tools to be scaled down to the size of cells and molecules, i.e. into the nanoworld. On this scale, drugs could be targeted so precisely as to exclude side effects. Genes and other molecular units could be repaired. Undesirable cells such as tumour cells and pathogens could be identified and eliminated directly.

Medical research has already made some progress in this direction, e.g. with miniaturised drug delivery devices, and with nanoparticles. However, developing a wide range of nanotech-based therapeutic approaches on the cellular scale remains a major challenge for interdisciplinary research between medicine, biology, chemistry and physics.

The book "Nanofabrication towards biomedical applications" maps the efforts to address this challenge. Structured with impeccable logic, it starts out with instructions on how to produce nanomaterials (five chapters), then tells us how to characterise them (four), how to use them in biomedical research (five), and what the impacts on society might be (two). Each chapter is a scholarly review of one of the aspects concerned, lavishly furnished with schematic diagrams, half-tone or colour images, and up to around 200 references.

So far so good -- as a collection of review articles, the book appears to work. But our expectations should be higher. In such an interdisciplinary field as nanobiomedicine, communication between specialists of different fields is one of the key requirements. When physicists write about what nanomaterials they can produce, a clinical researcher should be able to read the review and figure out how these materials might benefit her attempts to develop new therapies. Conversely, somebody writing about biomedical research should imagine an audience of physical scientists, in order to inspire them to invent new tools that might serve the medical goals.

This is where this book disappoints. Each of the chapters appears to be targeted at a narrowly defined specialist audience. To pick a random example, a section on "Optical detection of single molecules" *begins* with the sentence: "In confocal microscopy, coherent laser radiation is collimated and reflected off a dichroic mirror to fill the back aperture of a high numerical aperture objective." Try telling that to a clinician who is looking for new diagnostic tools.

The Borrowers

Biocatalysts and Enzyme Technology
K. Buchholz, V. Kasche, U.T. Bornscheuer
Wiley-VCH, 2005

Originally published in Chemistry & Industry No. 3 (6.2.2006), 29.

When it comes to developing new technologies, learning from Nature and borrowing natural materials is the oldest trick of the trade. Borrowing ready-made tools instead of fabricating one's own is particularly attractive in the realm of very small things, where our nanotechnology is still in its infancy and cannot match the performance of Nature's molecular machines. The mixed bag of many new and a few very old technologies based on the borrowing of cells and the macromolecules they contain is known as biotechnology.

"Biocatalysts and Enzyme Technology" is a textbook that deals with the catalysts that biotechnology -- and increasingly, organic chemists as well -- borrow from Nature, namely the enzymes (ribozymes are left out). The book, which is an updated translation of a German textbook published 10 years earlier, is structured in a logical and straightforward way, starting out from enzyme basics (protein structure through to Michaelis-Menten kinetics), then dealing with their production and purification, their applications, their immobilization and finally with the reactors and process technology.

The book comes across as a solidly constructed textbook, with exercises and ample references at the end of every chapter. The writing style -- readable, but quite academic, on the level of a review article in a scientific journal -- suggests it will reach more graduate than undergraduate students. Today's students, spoiled by so many full-colour textbooks, may find the illustrations disappointing, which include a few line drawings and micrographs (in greyscale), along with a surprising number of chemical formulae.

What is blatantly absent from the book, however, is an appreciation of the value of enzymes derived from extremophilic organisms. The issue is touched upon and misrepresented on page 84. In fact, intracellular enzymes of pH-extremophiles (acidophiles and alkaliphiles) don't need to be selected for pH stability at all, as the cytoplasmic pH of these organisms is no different from that of normal cells. In contrast, thermophilic enzymes are very much selected for their ability to function under extremely high temperatures, as microbes cannot insulate themselves from heat. The use of DNA polymerase from (hyper)thermophiles like *Thermus aquaticus* in the polymerase chain reaction is mentioned on page 89 as part of the recipe for PCR, but not appreciated as a prime example of biotechnology borrowing an extremely useful tool from the fringes of biodiversity.

I may be biased here, but it is my view that extremophiles have an enormous potential to provide enzymes that are stable under whatever extreme conditions a given application requires. Polymerases from thermophilic microbes and heat-stable lipases in biological washing detergents have already found widespread use, but with many industrial processes requiring extreme conditions such as heat, high pressure, high or low pH, or presence of toxic chemicals, there is a huge potential for enzymes from extremophilic microbes that might serve under these conditions. In my opinion, the authors should consider including a chapter about this area in future editions, possibly replacing the current chapter 7, which appears to transgress the boundary between enzyme and cellular biotechnology.

The big picture

Systems biology in practice: concepts, implementation and application
Edda Klipp, Ralf Herwig, Axel Kowald, Christoph Wierling, Hans Lehrach
Weinheim: Wiley-VCH, 2005

Originally published in Chemistry & Industry No. 6 (20.3.2006), 26.

The overwhelming success of biochemistry and molecular biology throughout the 20th century can be described as a process of dissecting biological systems into molecular pieces. The molecules of life were identified, then characterized in ever greater detail and in growing numbers, until the end of the century, when the electronic databases were bursting with detailed information concerning countless genes, genomes, proteins, and protein complexes.

The ultimate test of the accumulated knowledge, however, hadn't been done yet. Namely, if we put all the bits and pieces of information about the components of the living cell together again, do we get a coherent representation of a living, working cell? Does the wealth of details we have gathered really enable us to understand the whole?

Around the turn of the century, an increasing number of programs became available for the simulation of cells or major parts thereof, such as the e-cell, the virtual cell (vcell), and Gepasi. In June 2001, I reported on these early efforts without being aware of the new label "systems biology," which must have crept in around that time. As an idea whose time had come, systems biology established itself quite rapidly among the new interdisciplinary fields in the life sciences.

Now Klipp et al., five researchers from the Max-Planck Institute for Molecular Genetics at Berlin, attempt to define the nascent (inter)discipline with a textbook. At the systems level, the entire book presents itself as impeccably organised. The first part offers gentle introductions not only to systems biology itself, but also to the essential biology, mathematics, and experimental techniques. Depending on their background, readers can choose different paths to approach the subject and catch up on the basics they will need. The core of the book is made up of eight chapters covering specific aspects such as metabolism, signalling, gene expression, and evolution. Nerdy details of computer tools, databases and such like are confined to the third part.

The text is demanding, yet readable. Diagrams and equations are well presented. While the book would be too "hard" for lay readers, it should work well for ambitious undergraduates and graduate students. The five authors have managed to combine their skills and the very different biological and mathematical approaches required for systems biology into a homogeneous and meaningful whole.

On the IT side, the authors have demonstrated their computer skills by letting machines compile the references for them. Entire pages have been wasted on spelling out every single author on multi-author papers, like those reporting the sequence of the human genome (Lander et al., Venter et al.), not once but several times. Moreover, one reference listed on page 394 contains the text of the paper's acknowledgement section, which must have fused itself to the end of the title like a mobile genetic element. Clearly, no human eye has ever checked these references.

But such quibbles aside, the book seen as a whole system looks like a good starting point for all those who want to enter this young and promising field of life science research.

Nanotechnology sliced

Nanotubes and nanowires
CNR Rao and A Govindaraj
RSC Publishing 2005

Originally published in Chemistry & Industry No. 9 (1.5.2006), 25.

As scientific understanding of the nanoworld accumulates and our know-how in nanotechnology is beginning to catch up and develop into a respectable (inter)discipline with real-world applications, the fast-growing reservoir of scientific information spills over into the book market and drives a boom in nano-oriented (or at least nano-labelled) books, both academic and semi-popular. Among the various attempts at bundling nanotechnology between bookcovers, I have identified the Springer Handbook of Nanotechnology as the most convincing tome so far (Chemistry and Industry, 21.3.2005 pp. 24-25).

The arrival of Rao and Govindaraj's effort, however, throws up a new kind of consumer choice: Do you prefer your nanotechnology as a whole loaf or sliced? Because the present Nanotubes and Nanowires book (with 272 pages slightly thicker and much heavier than a slice of bread) is announced as volume 1 of a Nanoscience and Nanotechnology series edited by Harry Kroto, Paul O'Brien, and Harold Craighead. The publishers say they are hoping to release two volumes per year, with the following instalments slated to cover nanoparticles, fullerenes, and surfaces. So the question must be asked, is this a salami sliced too thinly, or is it the best thing since sliced bread?

As far as the presentation in text and pictures is concerned, the new book and the corresponding chapters in the Springer Handbook operate on a very similar level, and both are seriously academic yet readable. The sliced opus has around twice as many pages on carbon nanotubes and on nanowires, but the handbook squeezes more words into each page, so the overall difference should be small. A clear bonus for Rao and

Govindaraj, however, is that they offer a short but interesting chapter on inorganic nanotubes, which haven't got a chapter in the handbook.

The downside is that their volume is more than twice as costly on a per page basis than the handbook. At this price level, the yearly two volumes of the series will cost nearly as much as the entire handbook while covering only a fraction of the field.

Thus, it all boils down to the question how sharply focused or broadly extended your attention to all things nanotechnological will be. Anybody who needs to know everything about nanotubes and nanowires but not about the rest of nanotechnology will be happy with this book. Anybody who wants a compendium covering the whole of nanotechnology might find the handbook more affordable than the series. That is real consumer choice.

Nano future

The nanotech pioneers: Where are they taking us?
Steven A. Edwards
Wiley-VCH 2006

Originally published in Chemistry & Industry No. 20 (16.10.2006), 26.

Nanotechnology, once the realm of predictions and science fiction, has clearly arrived in the here and now. In their pursuit of ever smaller feature sizes, chip manufacturers have breached the 100 nm barrier, and thus brought their entire industry into the nanotech fold. Mundane everyday products such as cosmetics are increasingly based on designer-made nanoparticles. Medical applications of nano equipment including drug delivery devices and artificial sensory organs that connect to our nerves are under development.

But where will we go from here? And what will be the benefits, costs, and risks of going down the nano route even further? Steven Edwards has assessed the situation and outlook of the field from the perspective of a "technophile" with a keen interest in business aspects, who seems to spend more time over business and policy documents than with scientific papers. He has also done a lot of networking and interviewed a number of people involved in nanotech startup companies.

Accordingly, the resulting book is a bit weak on the deeds of academic pioneers such as Ned Seeman, Cees Dekker, and Charles Lieber, but very strong on reporting life at the sharp end of the nanotech business revolution. While most of the companies mentioned are in the US, the British Cambridge Display Technologies (CDT) and the French Flamel Technologies (named after a French alchimist who also figures in Harry Potter) also get a nod.

In my view, the most interesting chapter is the one on "grandiose challenges." Arguing that the field needs a focal point or three, the author picks three challenges that migh help to drive technology forwards, as the grand challenge of putting a man on the moon did in its time. His grandiose nanotech challenges are: Independence from fossil fuels, a space elevator, and a quantum computer of useful size. For each of these, Edwards outlines the state of the art, the most likely ways forward, and a timeframe by which he expects the challenges to be met.

Overall, the book is highly readable, even though the frequent references to the bible and some other American obsessions give it a distinctly US-specific flavour. All authors, myself included, make mistakes and depend on helpful friends and paid editorial staff to find them. Still, it is quite remarkable that the misattribution of the DNA structure to a certain "Joseph Watson" (p.91) not only survived all checks and edits, but even earned this fictional character an entry in the index, where it joins other mysterious phenomena, such as the "enome," and the split personality pair of "Al Gore" and "Gore, Al."

These things seem to suggest that (nanotech-based) robots are already beginning to take over the publishing industry. But the author appears to be a real human being, and his book is certainly worth reading.

Beyond evolution

Engineering the genetic code
Nediljko Budisa
Wiley-VCH 2006

Originally published in Chemistry & Industry No. 12 (25.6.2007), 31.

Protein biosynthesis is a central feature common to all cellular life forms. Over the past five decades, the sheer complexity of its structural and functional organization has confronted molecular biology with some of its biggest challenges. However, the field also promises some of the biggest rewards. A detailed understanding of how the linear genetic information leads to the self-organizing three-dimensional nanomachines known as proteins holds many tips and tricks that could become useful in future technologies, boosting our own attempts at conquering the nanoworld. Once we understand the mechanisms, we could also go beyond the range of molecular structures explored by nature and create entirely new nanomachines.

Apart from knowledge for the future, protein biosynthesis also promises rewards relating to our history. The task distribution between DNA, RNA, and proteins is one of the defining "inventions" in the early evolution of life. Transfer RNAs (tRNAs) are the adapters which in all cellular organisms hook up amino acids in order to match them to the correct codon on the messenger RNA. Moreover, tRNAs are also a unique window into the dark era beyond the life time of the last universal cellular ancestor (LUCA). As even that distant and primitive organism already had many different tRNAs along with tRNA synthetases to charge them, comparative studies of these molecules can point back to a more distant past when just one or two adapter molecules created the connection between a small number of amino acids and a much simpler genetic material.

How the genetic code evolved from its primitive beginnings, how and why it then stopped evolving (with a small number of exceptions), and how modern molecular biology can make it evolve further and broaden its range, that would make a compelling story ranging from the distant past into the near future, which in my view would be best told chronologically (although there are still gaps in our understanding of the beginnings). Making this story accessible to a general readership would require not only a firm grasp of the complicated facts of the matter, but also an exceptional clarity of writing, as there are many things that the reader could very easily get confused about.

When Nediljko Budisa turned his Habilitation Thesis at the Technical University of Munich first into a review in Angewandte Chemie, and then into this book, he must have aimed at reaching a somewhat wider audience than the immediate colleagues in the field who would have access to his papers anyway. He made a brave effort to link his own research, which aims at expanding the genetic code by introducing new functionalities, with the natural evolution of the code and of protein biosynthesis in general.

Sadly, however, the author does not master the clarity that would have been required to make the story outlined above accessible to a wider readership including, say, physicists and chemists. Thus, his audience will remain limited to the small circle of specialists, and the story of the genetic code from its very beginning through to its future remains to be written.

Speeding things up

Metal catalysis in industrial organic processes
Gian Paolo Chiusoli and Peter M. Maitlis (eds.)
RSC publishing 2006

Originally published in Chemistry & Industry No. 19 (8.10.2007), 32.

Catalysis is the most important facilitator of chemical reactions, no matter whether they occur in very small vessels like a bacterial cell, or in very large ones like an industrial reactor. Biology recruits its catalysts from just two groups of molecules: proteins (mainly) and RNA. Industrial chemists, by contrast, are spoilt for choice as they might in principle choose any substance as a catalyst as long as it speeds up the desired reaction and is safe to use. In numerous cases, finding the right catalyst is what turns chemistry into an industry.

A large group of man-made catalysts involves metals, either as solids in heterogeneous catalysis, or as organometallic complexes in homogeneous catalysis. The present volume gives an overview of metal catalysts, bridging the tradistional schism between homo- and heterogeneous catalysis. It is organised not so much by the specific type of catalyst, but by the type of reaction it is meant to speed up. Thus there are chapters by different authors on oxidation, hydrogenation, carbon-monoxide based reactions, carbon-carbon bond formation, olefin metathesis, and polymerization. The book is rounded off with a pair of appendices covering the fundamental science of homogeneous and heterogeneous metal catalysis. Eager students will find "discussion points" scattered throughout the chapters, designed to stimulate deeper thinking.

Each chapter provides a compact introduction to the relevant group of reactions and the catalysts used in their industrial applications. The metathesis of olefins, to pick an example that has seen some media attention following the 2005 Nobel prize and that is still good for new discoveries, is presented in a short and accessible summary. The chapter covers the different types (ring opening polymerization, ring closing, cross metathesis) and outlines both their industrial application and recent scientific progress.

The introduction, by contrast, is almost harder to read than the other chapters, as it tries to give a general and yet exact description of the economic and scientific importance of catalysis for industry. Thus, I expect that readers will pick and mix the chapters most relevant to their needs. Alternatively, they might want to look up a product of interest in the index and find out how it is made industrially, and with which catalysts. For example, there is a couple of pages on how to make menthol (not all of it comes from plants!) which I found quite enlightening. Only the most dedicated students aspiring to an industrial career will make their way from cover to cover.

Size matters

Nanoscopic materials: Size dependent phenomena

Emil Roduner

RSC publishing, Cambridge, UK, 2006

Originally published in Chemistry & Industry No. 23 (10.12.2007), 32.

Much of the textbook knowledge in physical chemistry is anchored in a fairytale world known as the bulk phase. In this remote realm, crystals are infinite in all directions, and no surface effects are allowed to disturb their perfection. Salts are dissolved at infinite dilution, and liquids know no boundaries. The complications that arise from the real world, where phases actually have boundaries, are often banished to the ghetto of surface science.

With the trend of recent decades to make things smaller and smaller, the bulk phase, if it ever existed, started to shrink into oblivion. In the world of nanoparticles, as Emil Roduner keeps reminding us in his book, there is no such thing as a bulk phase unaffected by surface effects. When you live on an island, all directions lead to the sea. And in nanomaterials, the surface is everywhere.

Accordingly, Roduner has redistilled the established knowledge of what used to be the (unjustly) marginalized field of surface science (which of course has important implications for many areas from bubbly drinks through to catalysis) into a form applicable to nanoparticles and other aspects of today's nanosciences. The book is based on his lecture notes for a graduate course he has held at the University of Stuttgart, Germany.

He starts with first principles including coordination numbers and electronic structures, moves on to thermodynamic and magnetic properties, as well as adsorption and phase behaviour of nanomaterials, and finishes off with two chapters discussing applications, first in catalysis, and then in the wider world.

For all its modern shift of focus, the book still reads like a mildly old-fashioned textbook of physical chemistry, requiring impeccable maths skills, concentration and possibly pencil and paper to keep up with the equations. There are lots of helpful illustrations, however, and some of them in full colour. And if all else fails, each of the eleven chapters includes a summary box at the end, where the "key points" are reiterated in clear and concise language, complete with bullet points.

While it's too hard to reach the famously elusive intelligent lay reader, this book provides a useful introduction for students who approach nanomaterials and nanotechnology from a physical chemistry background, e.g. as a late undergraduate or graduate optional subject. To the rest of us, it is a timely reminder of how much everything we think we know about matter can change when the packaging unit of that matter becomes very small.

Not a mirror image

Asymmetric synthesis: The essentials
M. Christmann, S. Bräse (eds.)
Wiley-VCH 2007

Originally published in Chemistry & Industry No. 5 (10.3.2008), 29.

Recently, the thalidomide tragedy of the 1960s has been making headlines again – with the showing of a TV drama in Germany, which Grünenthal tried to stop, and renewed arguments over compensation. To the chemist, of course, this is also a stark reminder of how important it is to produce the correct enantiomers of biologically active substances.

In the cell, the enantiomeric purity of all biomolecules is so perfect and ubiquitous that it hardly ever gets mentioned. I can think of only one paper that investigated the mirror image of a protein. In the lab, however, getting the chirality right is still a huge challenge, which grows even bigger as the chiral compounds that chemists try to synthesize get more complex. Given the importance and the size of this challenge, it is surprising that there is relatively little help in the book market on how to get the right enantiomer.

The book by Christmann and Bräse, with the title promising "the essentials" of asymmetric synthesis, might be mistaken for an introductory textbook. Closer inspection, however, reveals that it is in fact a collection of invited minireviews published to celebrate the 60th birthday of Dieter Enders, who has made many influential contributions to the field since the 1970s. As always with such volumes, the content is determined by network structures more than by what the audience needs to know or hopes to learn.

The contributions from over 80 researchers are grouped in five major sections, on chiral auxiliaries, metal catalysts, bio- and biomimetic catalysis, total synthesis, and industrial applications. Each section begins with a more general chapter designed to give an introductory overview.

The chapters abide to very strict length regulation: except for the introductory overviews, none exceeds 6 pages including references and CV of the authors. The highly condensed format means there is not much room for didactic help, so many of the chapters will leave non-specialists dazed rather than enlightened. Then again, the compact format of the contributions gives challenged readers (including this one!) hope, as even the most incomprehensible chapter will end soon, and a different topic will come up right behind it.

All in all, the book appears to be too dense and difficult to serve as an introduction to the field. It will be useful, however, as a research guide. For instance, graduate students in organic chemistry who are beginning to look for a postdoc placement, may find this volume a useful place to start browsing. So the book is really a help to find the right lab or person in asymmetric synthesis, rather than an instruction how to get the right enantiomer.

Science goes bonkers

Bonk: The curious coupling of sex and science

Mary Roach

Allen Lane 2008

Prose and Passion 4.7.2008

I have by now accumulated a nice little collection of books about the science of love and sex -- of course for serious research only. On that shelf, "Bonk" is by far the most entertaining book I've come across. Not worrying too much about love and all the emotional baggage that we humans connect with it, Roach goes, chimp-like, straight to the hard core of the matter.

Which of course is a very tricky thing to do for somebody writing in English, the language of puritanism and all kinds of post Victorian hang-ups around human sexuality. Notwithstanding, Roach attacks the matter fearlessly and head-on, using her fine sense of humour to overcome any embarrassment she or some of her readers may be feeling.

The book is mainly about how scientists manage to survive in a field that suffers enormously from the above-mentioned hang-ups and prudishness and is therefore still the most under-developed part of human biology (remember that a large part of the clitoris was only discovered a few years ago!). One doesn't get much of an overview of what is known about love and sex (for this, the book "Lust and love: is it more than chemistry?", which I translated, works better, even if I say so myself), but one does get a good sense of the challenges -- resulting from nature, methodology, and from the weird attitude our society has towards the most natural of all inter-human activities -- and how the very small number of researchers in the field are facing them. As a bonus, one gets to LOL a lot.

A plodding life

Max Perutz and the secret of life

Georgina Ferry

Chatto & Windus 2007

Prose and Passion 5.8.2008

During my week off, I read Georgina Ferry's biography of Max Perutz, who solved the crystal structure of haemoglobin in an epic quest lasting over 3 decades, and who set up the Laboratory of Molecular Biology in Cambridge, which I understand has produced more Nobel laureates than France or Canada in the time since it opened.

All this makes for an exciting story, even though the protagonist is anything but a glittering star. Perutz was a very patient and persistent "plodder" who, while eager for success and recognition, was never seen as a genius and never had the over-sized ego that often comes with such a label. The persistent plodding won him the haemoglobin structure and the Nobel prize, while his modesty allowed him to quietly run a world-leading institute where he had to handle primadonnas like Francis Crick.

Obviously, the book is a must for anybody interested in proteins. For everybody else, I was worried a bit that it might turn out a bit boring as I knew that Max was a less than glittering person. But I think the author has managed the trick to turn his plodding life into a compelling story, which should be interesting for non-specialist readers as well. The main lesson for the general public is, of course, that one doesn't have to be a towering genius of stature of a Crick or Bernal in order to be a successful scientist. Relatively ordinary people can make an impact too.

Digital takeover

The chemistry of photography: From classical to digital technologies
David Rogers
RSC publishing, 2007.

Originally published in Chemistry & Industry No. 7 (7.4.2008), 28.

Last year, my teenage daughter decided to reactivate the dark room equipment that has been in our family for generations, so I went to a well-known photo retailer on the high street to buy the photo paper and chemicals required. The middle-aged shop assistant looked at me as if I had climbed out of a UFO crash-landed in his window. Only after I repeatedly assured him that I had spotted the desired materials on the top shelf of this very shop a couple of years earlier did he endeavour to go and look for them. He did find all that I needed eventually, but not without informing me that in his many years of working in this place, I was the first person to ask for such things.

So is the traditional, wet chemistry way of producing photos dead already? If it is, the publication of a brand new book on the chemistry of photography looks a little bit odd. Who will need a new book on a disappearing technology? The idea seems to have been to summarise the state of the art before the art disappears. Maybe as a time capsule for future generations.

The author, who doesn't reveal much about his personal connection with old-style photography, spends 13 chapters on the details of photographic papers, films, and processing technologies, relying mainly on the patent literature to summarise the way things are done in the industry. There are colour illustrations throughout, and lots of chemical formulae, too, so the presentation appears to address a highly specialised, professional audience, which one suspects is dying out along with the technology.

Patched on to the end of the book we find a 14th chapter that explains how inkjet paper works. This appears a bit thin, considering the grand promise of "digital technologies" in the title. I'm sure there must be some chemical aspect to the sensing of coloured light in digital cameras, and to its reproduction in data projectors and various types of printers.

I think this book really is a time capsule. It will be most useful for historians of the 22nd century, who will rely on it to find out about extinct technologies of the 20th century. I'm afraid that for young enthusiast like my daughter who are trying to keep the technology alive against the odds it isn't very useful. And the people in the industry will know these things already and will not have that many new staff members to train. So my advice is to wrap the book up and store it in a safe, dry place for a century or two, it may come in handy one day.

Put another nickel in

Nickel and its surprising impact in nature (Metal ions in life sciences, Vol. 2)
A. Sigel, H. Sigel, R. K. O. Sigel (eds.)
Wiley 2007

Originally published in Chemistry & Industry No. 24 (22.12.2008), 29.

In geometry, cutting across a three-dimensional shape can produce surprising features, such as the circles, ellipses, parabolas and hyperbolas hidden in the innocent-looking double cone. Similarly, cutting across the established structures of scientific knowledge can lead to surprising insights.

The present monograph, the second in a new series on bioinorganic chemistry, cuts across a wide range of biological topics, using the element nickel as the scalpel that exposes surprising connections and fresh surfaces.

One of these sections concerns the pathogen *Helicobacter pylori*, which we now know to be responsible for gastritis and a range of follow-up problems from ulcers to cancer. This notorious bug survives in the stomach by neutralising its local environment with ammonia, which it produces with the help of urease, a nickel enzyme.

Urease, in turn, cuts across the history of science in a rather intriguing fashion. Jack bean urease was the first ever enzyme to be crystallized (1926), although the first urease crystal structure arrived rather late (1995), and the structure of the pioneering jack bean enzyme was only solved in 2002. Urease was also the first enzyme that was proven to depend on nickel for its activity (1975).

In plants and animals, urease resides in the cytoplasm of the cell. *Helicobacter* is the only kind of organism known to secrete the enzyme into the environment. Therefore, infection with *Helicobacter* can now be tested non-invasively with a rapid urease test, and the enzyme is also an attractive drug target. Helicobacter mutants lacking the enzyme have been proven unable to colonise the stomach.

For all these reasons, this book has dedicated chapters on urease (chapter 6) and on *Helicobacter* (chapter 15), and the enzyme also appears frequently elsewhere, e.g. in chapter 5, which is about active site models for nickel enzymes. There are six other classes of nickel enzymes known today, each represented by a chapter. The rest of the book deals with general issues such as transport and homeostasis (regulation of nickel concentration), interaction with biomolecules, and gene regulation in response to nickel. The final two chapters deal with the impact on human health, i.e. nickel allergies and toxicity.

The biological role of nickel isn't one of the boxes that science gets filed into; I suppose there aren't any departments dedicated to nickel biochemistry. All the more interesting then, to use this metal to slice across the established classifications and think outside the box for a while.

Nature's chemistry kit

The ubiquitous roles of cytochrome P450 proteins (Metal ions in life sciences, Vol. 3)
A. Sigel, H. Sigel, R. K. O. Sigel, eds.
Wiley 2007

Originally published in Chemistry & Industry No. 21 (10.11.2008), 30.

Anybody with a general interest in proteins will have come across cytochrome P450 proteins in one way or the other. I must admit that I've often stumbled over them and never managed to make much sense of what I read about them, as each example I encountered seemed to be entirely different from the last one I vaguely remembered. So what are these proteins for, why do they have such a weird and confusing nomenclature, and why are there so many of them?

Fortunately, this book answers all these questions and many more. The P450s are also known as monooxygenases, because they only use one of the atoms of the oxygen molecule as a terminal electron acceptor (which ends up in a water molecule). The other oxygen atom – and that's the secret behind the astonishing chemical diversity of these proteins – can be used to do chemistry.

Depending on which monooxygenase produces it, this leftover oxygen atom can react with hydrocarbons, halogens, alcohols, aldehydes, ethers, thioethers, amines, nitriles, and N-oxides. What we find here is Nature's chemistry set, which can catalyse thousands of different oxidation reactions. However, looking at the bigger picture, we find that these reactions can be assigned to three major biological functions.

The first is degradation of unwanted organic molecules, which today is relevant because it includes the degradation of pharmaceuticals. Then there is a wide range of synthesis reactions serving the production of various hormones. And finally, the transformation of fatty acids.

This chemical diversity also explains why there are so many P450s. They exist in all eukaryotes and archaea, an in many types of bacteria. With the rapidly growing bounty of genome sequences coming in, we can also look at how many different P450 genes each species has. We humans have 55, which might look impressive until you consider other species. The mosquito *Anopheles gambiae* has 100, and the rice plant 356 such genes. Plants in general have lots of P450 genes, which can account for up to 0.6 % of their genomes.

What looks terribly confusing for the layperson has proven a unique treasure trove of molecular biology and structural biochemistry. There is no comparable example of a single protein family, clearly arising from a single ancestor, but diversifying into thousands of different enzymatic functions.

To keep track of all this, P450 researchers have from very early on developed a naming system specifically for these proteins. Using certain thresholds of sequence identity, they divided the proteins into families (CYP1, CYP2 ...), then subfamilies (CYP1A, CYP1B ...), which are further divided into numbered subsets.

With its very readable introduction and 16 further chapters on specific topics in the structural and molecular biology of P450s, this monograph offers even those who were previously scared off by the confusing multiplicity a door into the fascinating world of a uniquely diverse family of biological catalysts.

Genome FAQs

A short guide to the human genome

Stewart Scherer

Cold Spring Harbor Laboratory Press 2008

Originally published in Chemistry & Industry No. 13 (13.7.2009), 30.

We live in the post-genomic era, so the human genome is, supposedly, fully brought to book and neatly filed in searchable databases. However, for anybody who is not directly involved in genome research, and therefore not familiar with the database tools needed to "read" a genome, it can still be extremely difficult to get sensible answers to seemingly very simple questions.

For instance, if I as a science writer want to report that the genome sequence of the house cat has been published, and 18,925 genes have been identified, the obvious thing to do would be to compare this number (which I just made up) to the number of genes in the human genome. But how many are there?

I remember the pre-genome times when the estimates ran from 50,000 up to 100,000. As the genome sequenced neared completion, they dropped below 30,000, and are now closer to 20,000. But to this day, researchers cannot give a straight answer to this simple question.

In his "short guide", Stuart Scherer has addressed around 80 such questions, using the available database tools and the published literature to give today's best estimates, pinning the genome and its proteins down as precisely as possible. Regarding the number of genes in our genome, his answers range from 18,357 to 25,685.

Each of these "genome FAQs" is addressed in a very short chapter of just one to four pages, often presenting the answer in the form of a table or graph, together with an explanation of how the results were obtained and what factors still limit their precision.

One interesting example is the size of proteins encoded by the genome. I would have expected a curve rising to maximum at around 200 amino acids, then fading away in a long tail towards the small number of extremely large proteins such as titin, which functions as a rubber band in muscle.

The graph which Scherer displays has a few additional features. While there is a noisy kind of local maximum around 200, there is a second bump around 300 amino acids, culminating in a spectacularly sharp peak at 312 amino acids. Apparently we have 79 proteins which have exactly this number, plus 105 which have one more or one less. Large protein families, such as those involved in the sense of smell, offer a partial explanation to this remarkable feature, but it still surprises me.

Other questions covered in the book range from the very basic, e.g. frequency of nucleotides and dinucleotides in the genome and in individual chromosomes, sizes of chromosomes, gene density, etc. to the more complex issues including gene families, mobile elements, and genetic variation between individuals. The book closes with a chapter on comparative genomics, i.e. what we can learn from comparing our genome to those of other species.

This is only a slim book, but it is a very useful reference for everybody looking for information about our genome.

Shrouded in mystery

Titan unveiled

Ralph Lorenz and Jacqueline Mitton

Princeton University Press 2008

Originally published in Chemistry & Industry No. 16 (23.3.2009), 30.

If chemists were to choose the destinations of space exploration there would be one target that would get the highest priority: Saturn's moon Titan. Its atmosphere is the most Earth-like that has so far been discovered. With 1.5 bars at ground level, over 90% molecular nitrogen, and the rest being mostly methane, it is very close to what our atmosphere must have been like before oxygen came along. There is a whole range of organic molecules to attract a chemist's attention, which make up the characteristically orange "haze" that shrouds Titan's surface. There may be methane rivers and we now know for sure that there is at least one lake containing liquid ethane. Outside our own world, you can't find that much chemistry anywhere else in the solar system.

Following the flybys of the Voyager probes in the 1980, which produced spectacular images but did little more than tickle our curiosity regarding Titan, the US and European space agencies NASA and ESA devised a joint mission to the Saturnian system comprising the orbiter *Cassini* and the lander *Huygens*. The mission arrived at Saturn in June 2004. *Huygens* made a successful touchdown on Titan on January 14, 2005, while the mothership *Cassini* completed its four-year mission and is now well into an extended two-year mission with particular attention to the moons Titan and Enceladus.

The story of Titan exploration clearly deserves to be told in a popular science book (or even in several books), and *Cassini* veteran Ralph Lorenz and astronomy writer Jacqueline Mitton have had a good go at telling it from the very beginning to the mid-point of the primary mission, which is June 2006. The chronological account is lightened up by "blog entries" giving Lorenz's experience at the research front line first hand. The most memorable of these describes how, in 2005, he retrieves a computer program he wrote for the *Huygens* landing some time in the early 90s. As he goes through the motions of firing up a half-dead PC, we realise just how much time it takes to get a mission to the Saturnian system.

Timing, however, may also be the book's downfall. Although it only came out in 2008, scientific results are only included up to June 2006 and the production deadline was December 2006. Thus, at the time of reading it, as *Cassini* has started its extended mission and ethane lakes have finally been confirmed, it already looks dated. Seeing that the same authors already had a book out on Titan covering the situation prior to the arrival of the mission, one wonders why they couldn't wait until the completion of the primary mission, and maybe persuade their publishers to move a bit faster. While travel to Saturn necessarily takes years, publishing a book doesn't have to take that long.

Why music doesn't add up

How equal temperament ruined harmony (and why you should care)
Ross W. Duffin
Norton paperback 2008

Prose and Passion, 7.1.2008

Music, I was led to believe, is a supremely elegant manifestation of pure mathematics. The intervals we know as fifths (think "Twinkle - twinkle"), fourths (Auld lang syne), and octaves (Somewhere over the rainbow) correspond to simple fractional relations between the sound frequencies, of 2/3, 3/4, and 1/2, respectively. And as an illustration, one gets shown the corresponding keys on a piano keyboard.

What nobody told me in the first 44 years of my life is that the intervals you play on the piano do not correspond to the simple fractions cited above. I only started hitting on this problem when my daughter's cello teacher dropped a hint that the piano was out of tune by default. The piano is actually tuned not in pure intervals but in a system called “equal temperament” for the simple reason that the fractions don’t add up. If you add up 12 fifths, all around the circle of fifths (C - G - D - A - E - B – F# - C# - Ab - Eb - Bb - F - C) you get $(3/2)^{12} = 129.746$. Theoretically, the first and the last C in this series should be seven octaves apart, so their frequency relation should be $2^7 = 128$. And not 129.746. And there are even worse clashes with other intervals. So in fact it would be impossible to tune a piano according to the pure intervals defined by simple fractions. This is why we as a civilisation have settled for equal temperament, which means the octave is split into 12 equal semitones.

Equal temperament (ET) is so widespread today that knowledge of the alternatives has gone missing, and even many musicians are unaware of the problems that this compromise solution causes. Duffin argues that some of the “unequal” solutions favoured in renaissance music and through to the end of the 19th century (he dates the total victory of ET to 1917) would still be useful today and that the question of temperament should be considered afresh for each piece of music, taking into consideration the likely intentions of the composer, the context of its creation, and what’s best for its harmonies. This will all be self-evident for practitioners of early music who use historic instruments and temperaments already, but it may be new to many people dealing with the classical repertoire from Bach to Beethoven (who, the author argues, cannot have become used to hearing ET by the time he went deaf).

This argument is all very well and convincing, but it would fill only around 30 pages, so to bulk his pamphlet up to a marketable 196 pages, the author has included lots of repetition (as you tend to do in music!) and biographical profiles of everybody who has ever voiced an opinion on temperament, from Mozart’s father, via the flautist Quantz, through to the cellist Pablo Casals. And cartoons. And diagrams. But all this is redundant in principle, so if you’re just after the meat of the matter, you can probably read the relevant pages within an hour, at a bookshop cafe.

What remains is the impression that music is in fact a lot less mathematically elegant than it is often claimed to be, and that it is a rather messy compromise between pure mathematical beauty and practicability. The good news is that a messy system leaves you free to mess with it, giving performers more freedom. So from now on, when I play out of tune, I can always claim I am experimenting with different temperaments.

Silly questions

Can cows walk down stairs?

Do cats have bellybuttons?

Paul Heiney (ed.)

Sutton Paperback 2006 and 2008, respectively

Prose and Passion, 13.1.2009

These two books came out of a remarkable institution that lived only for a few years. London-based ScienceLine (not related to the current project scienceline.org) offered scientific answers to any question that ordinary citizens chose to fire at it via phone or email. After a few years and 16,000 questions answered by 8 gurus (teachers? scientists?), government funding for the project ceased and it closed down in September 2003.

These two books are its legacy – two colourful bouquets of vaguely science-related questions plucked from the ScienceLine archives and arranged by non-science author Paul Heiney.

I came across them because the youngest member of my family picked up the first of these books and enjoyed it. With the style of the questions very much along the lines of children's questions (we are told nothing about the age or demographic profile of the people who asked them! They may be mostly children, for all I know.), the short(ish) answers and the funny cartoons, it's no wonder the books appeal to children, and probably also to grown-ups whose scientific education got stuck at that level.

For the scientist, they are a mixed blessing. Reading most of the questions but only a selection of the answers, I found equal measures of interesting and trivial stuff on both sides. Some of the answers contained silly errors that should have been eliminated at proof reading, e.g. mix-ups of the effect of convex mirrors (in book 2, the answer suggests wrongly that it makes things appear closer). Some answers didn't match the question (one question was about the limits of the galaxy, and the answer was about the limits of the universe), and some left me unsatisfied, including the answer to the title question regarding cows and stairs. Now I know that they can walk upstairs but not downstairs and that it has something to do with their knees, but I still don't understand why they can't walk down. I'm sure cows in the Swiss Alps can walk down the mountainside!

The questions are a reminder of how difficult it is to ask a good scientific question. If these are the 500 best ones, I feel for the people who had to answer the 15,500 questions that were not selected.

Still, the kind of dialogue established by ScienceLine is a good idea in principle and it's a shame it was discontinued. If I were to establish something like this, I would probably avoid promising to answer every question. If, say, only the 5 best questions get answered every day, and one of them is syndicated out to be printed in a few newspapers for a fee, maybe ScienceLine could even be commercially viable without that government grant that proved so short-lived. At least the number of questions sent in and the popularity of the books suggests that there is a demand for this kind of dialogue.

PS: A nitpicker's guide to some of the other errors:
Asked whether one is more likely to win the lottery twice in a row or twice in a lifetime, the gurus say "it would be easier if" and go on to answer a completely different question. In fact the question is very straightforward, as the outcomes representing two wins in a row are a subset of those for two in a lifetime. Thus, anybody playing the same lottery more than twice in their lifetime is more likely to achieve the latter result than the former.
Asked how many prime numbers exist, the gurus give the right answer (infinitely many) and proceed with a wrong explanation, saying it must be infinite because the whole numbers are infinite as well. In fact, there is a very simple proof that would have enlightened the readership:
Imagine the opposite were true, and there were only a certain number of primes. Write them all down, multiply all, and add 1 to the product. Call the result P. Try dividing P by all known primes. It is not divisible by 2, as there will be a remainder of 1. Same

situation for 3, 5, 7, and all primes on our original list. Hence P is a prime that was not on our original list, proving our initial assumption wrong.

Balancing biological resources

Biocatalysis and bioenergy
Ching T. Hou, Jei-Fu Shaw (eds.)
Wiley 2008

Originally published in Chemistry & Industry No. 10 (25.5.2009), 30.

Biofuels were all the rage when this book appeared in 2008. Large-scale programs using fuel crops expanded rapidly, spurned on by petroleum prices climbing to unprecedented heights. Meanwhile, concerns over food prices and environmental impact of intensive farming of fuel crops inspired efforts to develop alternative, so-called second generation biofuels based on agricultural waste materials or crops that don't compete with food crops.

By the end of the year, however, a significant drop in the price of crude oil, along with the pressing problems of the financial and economic crisis, pushed biofuels off the agenda and the front pages. Still, the problems that biofuels are meant to address aren't going away either, and it is reassuring to know that there are many people pressing on with research towards better fuels, come rain or shine.

This monograph, emerging from a conference held in Taiwan in 2006, assembles reviews mainly on bioethanol, biodiesel, and other products made either with biocatalysts or from renewable resources. The latter comprise about half the length of the book, leaving a third of the space for diesel and just a sixth for bioethanol.

In marked contrast to the public debates over biofuels, food, and climate change, these research reviews are very much from the grassroots of the field, giving even the smallest technical problems due attention. It is impressive to learn from the first few chapters how many parameters must be controlled before plant material can be used as diesel fuel. Ignition temperature and melting point are obvious things that one must get right, but each type of fuel also has its characteristic profile of exhaust gases (apparently, biodiesel tends to produce less particles, hydrocarbons, and carbon monoxide than petrodiesel, but more nitrogen oxides) and lubricating properties.

Represented by the shortest part of the book, bioethanol gets less attention than it probably deserves, given that it is being used on a very large scale already (e.g. in Brazil and in the US), and that a switch of its production from first generation methods to second generation ones might produce substantial benefits.

In the second half of the book, with its diverse collection of things that can be made from natural sources, one can learn a lot about natural resources that have so far been underused or underappreciated. There is a strong delegation of researchers interested in marine resources, so I found out about the untapped potential of lipids from seaweeds, and how many megatons of fish caught are discarded unused every year, even though they could be turned into useful products like bioactive peptides. Considering the current threats to fish species like tuna and cod, it is intriguing to learn that there are so many other marine resources that are underused. We just need to find the right balance.

Every spin-out tells a story

Spin-outs: Creating businesses from university intellectual property
Graham Richards
Harriman House 2009

Originally published in Chemistry & Industry No. 12 (22.6.2009), 29.

Oxford University has been a seat of learning for many centuries, but only became a hive of commercial activity in recent decades. Although it had a very successful start with the foundation of Oxford University Press in 1478 (which was never spun out and continues to thrive under the somewhat archaic management of the university), it missed out on some major opportunities in the 20th century (penicillin, cephalosporins), and only really got going in the 1990s.

Since 2000, the university's very own technology transfer office, Isis Innovation, has spun out an average of seven companies per year, and the university's chemistry department secured £ 20 million in private funding in exchange for a cut of the university's share in any spin-outs created from the research at that department for a fixed period of time.

Graham Richards played a key role in both these radical changes. As a researcher in theoretical chemistry, he was the academic founder of the fourth spin-out company, Oxford Molecular, he played a role in making the university's technology transfer more efficient, and as the first chairman of the newly united chemistry department, he fixed the deal which secured the new laboratory.

In this little book, part memoir, part advice to future academic spin-out founders, he very briefly sketches his involvement with Oxford Molecular and with Oxford's wider technology transfer activities, along with the briefest of mentions of related activities in the UK, such as the British Technology Group, later BTG.

As somebody who worked at Oxford's chemistry department in the moment it was born, and who already had some vague ideas of the Oxford Molecular story and memories of some of the other people and companies mentioned, I enjoyed reading this account and piecing my fragmented impressions together to make a meaningful story. Similarly, any academic involved with research that might be suitable for a spin-out will surely find inspiration and some basic advice in these pages, even though they are not comprehensive enough to serve as a "complete idiot's guide to setting up spin-outs".

The opportunity that the author missed, however, is to turn the book into a more widely appealing memoir that everybody could enjoy and learn from. There are glimpses of a great story here (Oxford Molecular co-founder and CEO Tony Marchington with his natural business instinct and passion for steam engines appears to be a perfect character to write books about), but they aren't really exploited. I reckon that every spin-out company tells a story, but only a few of them have been written up so far. Most notably, back in 1994, Barry Werth told the spellbinding story of Vertex in "The billion-dollar molecule". Maybe, now that he has retired, Richards will find time to tell the whole fascinating story in a larger book, steam engines and all.

Collision course

Collider: The search for the world's smallest particles
Paul Halpern
Wiley 2009

Originally published in Chemistry & Industry, No 23 (7.12.2009), 27.

On September 10, 2008, the first proton beams entered the circular racecourse known as the Large Hadron Collider (LHC) buried deep in the ground under the French/Swiss border near Geneva. The record-breaking instrument had been launched with the promise to explore new physics, to mimic the conditions at the origin of our universe, and to find the elusive Higgs boson, believed to confer mass on other particles. It also attracted fears that unexpected phenomena might lead to disaster. What happened instead was that, within the second week of operation, the instrument developed a serious fault, and the repairs took nearly a year.

The delay in LHC's operation allows us non-experts time to catch up with trying to understand why physicists want hadrons to collide, and what they are hoping to learn from the fallout of such collisions, with expert guidance from physicist and author Paul Halpern.

Halpern, whose previous works include "What's science ever done for us?" (reviewed in Chemistry & Industry, No.2, 28.1.2008, 29), tells two intertwined stories leading to the much anticipated LHC experiments. One is the evolution of knowledge about fundamental particles, progressing from atoms, to protons, to quarks and strings. The other is the evolution of particle accelerators and colliders, which conversely tends to produce larger and larger entities as the size of the particles studied shrinks.

Explaining particle physics to lay readers is a daunting challenge, not only because it's a difficult subject matter, but also because every reader will have dropped out of the learning curve from atoms to strings at a different point. For me, like for many chemists, classical quantum mechanics still makes sense, while everything from Feynman onwards appears a bit fanciful. But other readers need to be picked up at different points, which is difficult to do in book format without boring one part and over-stretching the other part of the audience.

Halpern addresses this challenge with generous helpings of metaphor and explanation, spiced with the occasional pun. Yet, as a writer facing similar problems, I wonder whether those readers who need to be told what protons and neutrons are will stay the whole course through to the point where they can appreciate LHC experiments.

What I found most exciting in this book, however, was the second story, of how physicists came to build bigger and bigger accelerators, ending up with the LHC. There is much to be learned about the nature of human endeavours from this side of the book alone. For example, the fate of the Superconducting Super Collider (SSC), which was to be built from scratch in a remote part of Texas, until Congress pulled the plug, contrasts with the LHC which is based on using existing CERN infrastructure and recycling as much as possible from previous instruments, and thus could be financed without much pain.

Physicists have to think big to find the smallest particles, but they also have to think international, sustainable, and fundable. CERN and the LHC show how it can be done.

III. Literary reviews

“Literary” may be not be the right label for all of these, but this section contains reviews of some of the other books I’ve read, apart from all that science-related material reviewed above. All of these reviews come from my blog, which only began in 2006, so this section cannot claim to be a representation of the decade, and not even of the books I read during the decade. But some of these texts may still be worth keeping, so I just threw them in for good measure, which leads us directly to the first title ...

Measuring the World

Die Vermessung der Welt
Daniel Kehlmann.

Prose and passion 22.3.2008

Measuring the World is a somewhat fictionalised double biography of Carl Friedrich Gauss and Alexander von Humboldt. In their parallel lives, both did indeed measure parts of the world, one as a mathematician and surveyor on home ground, the other as an explorer and naturalist.

The book is very readable and follows in the tradition of books like "Longitude". This "mid-brow" terrain is still relatively new to German literature; until recently there was a wide canyon separating the U and the E literature, i.e. the popular and the literary fiction. Novels bridging this gap, possibly starting with Patrick Suesskind's "Perfume" have been rewarded with successful translation deals, and Kehlmann's book is now available in English too.

While it was very entertaining to read, and useful in terms of organising a number of household names of the 19th century into a network of who knew whom (they didn't have facebook back then!), I felt slightly unsure about the fine line between biography and fiction. I mean, did Gauss really make that promise to learn Russian as a favour to a Russian prostitute? I'll have to read a proper biography at some point to clarify this important point.

The artist inside

El cristo feo
Alicia Yanez Cossio

Prose and passion 7.4.2008

Michelangelo famously stated that his statues were already present in the marble a priori, and the hand of the sculptor did nothing but to liberate them. Although the author doesn't mention it directly, this may well have been the inspiration behind "El cristo feo" by Alicia Yanez Cossio (Quito, 1929). All the main figures in this novel, namely a statue and three living persons, get re-sculpted in different ways.

This is the story of the poor maid Ordalisa who works for a zombie-esque elderly couple. She has inherited an excruciatingly ugly crucifix from her mother, which one day starts talking to her. It (or Christ, or whoever is talking, it might just be within her head!) has a refreshingly unorthodox take on christian values, reminiscent of the 1970s cult hit "Mr God this is Anna". Thus, even if the story is superficially about a crucifix, it's never preachy.

Instructed by the Voice (who is and isn't inside the crucifix, as s/he is and isn't everywhere), Ordalisa starts chiselling out a beautiful (and movable) christ from inside the ugly crucifix. In doing this, she also liberates the artist within herself, and a living person within one of her employers (although the other gets resculpted in a less commendable way).

Told laconically with a minimal cast and setting, this enchanting story would work just as well as a chamber piece for a small theatre. Michelangelo would have loved it, as will anybody who has a bit of an artist hidden somewhere underneath their marble surface.

Going loco

El pergamino de la seducción (The scroll of seduction)
Gioconda Belli

Prose and passion 25.11.2008

"Juana la loca," or Queen Joanna of Castile never reigned because her father, her husband, and her son consecutively declared her insane and unfit to govern, claiming her political role for themselves. According to Gioconda Belli's rehabilitation dressed up as a novel, the poor woman wasn't insane at all, just squeezed between three males who had much bigger political ambitions and instincts, and whom she had the misfortune to love. (The men in question were Ferdinand II of Aragon, her father, Philip the Handsome, her husband, and her son Charles I of Spain and V. of Germany, in whose empire the sun never set.)

Gioconda Belli (born 1948) brings Joanna back to life using Lucia, a young woman in 1960s Madrid (so she's vaguely of the author's generation, who also went to boarding school in Madrid in the 60s) who plays her role in a game of history re-enactment that allows her to go through at least some of the experiences (love, vulnerability, imprisonment, bereavement) that Joanna went through in the 16th century. The role play artifice bridging a gap of 450 years works surprisingly well, bringing both the unfortunate queen and her modern day interpreter to life. The male history prof who sets the situation up and manipulates Lucia as the three men in her life manipulated Joanna remains more enigmatic. The spooky old house where much of the story is set is almost a character of the story and reminded me of the modern classic Nada by Carmen Laforet.

Does the author convince us that Joanna wasn't mad after all? In her afterword she argues that while historians tend to label her as schizophrenic, none of the psychiatrists she consulted agreed with this diagnosis. Most tellingly, Joanna only went "mad" when she was treated badly, e.g. locked up, forcefully separated from her children, etc., but appeared perfectly sane under normal conditions. Which suggests that she wasn't psychotic at all.

What appears to have been her real problem, however, was a lack of political instinct, which turned out to be catastrophic for somebody thrown into the political arena simply by virtue of her royal descent and her robust health – everybody else with higher ranking claims to the throne died young. Yes she was kicked around by the men in her life, but one can't help thinking that the queen gene skipped a generation. Her mother, Isabella of Castile, described as cold-hearted in the novel, but famous for the Reconquista (the expulsion of the Arabs from the Iberian peninsula) and for sponsoring Columbus' crossing of the Atlantic, would have asserted herself in a similar situation, while Joanna just drowned in it and went "mad".

Her gift appears to have been in a different domain – she is described as a passionate lover (the safe hands of an author who is also known for her erotic poetry are a definite bonus in that domain) and managed to produce six healthy children who all went on to become monarchs. The mad world of political power-struggle just wasn't for her.

Culture clash

K – The art of love
Hong Ying

Prose and passion 5.12.2008

K is the story of a strange encounter between two cultures. At the surface, it's about Chinese and English culture, and also about the very Martian culture of a man and the Venusian one of a woman. But it soon becomes clear that the protagonists aren't representing these cultures. If anything, they struggle to define an identity within small subcultures at the margins of their respective societies.

The Englishman, Julian Bell, is like his eponymous real-life model a product of the Bloomsbury Group which had a set of values quite radically different from what was considered normal at the time. Son of the painter Vanessa Bell (who had an open marriage with a bisexual man) and nephew of Virginia Woolf, he tends to judge everything with the measure of the intellectual cult he grew up in, and initially sneers at the idea that Chinese poets may be producing anything comparable.

The Chinese woman, called Lin Cheng in the novel, but based on the biography of the poet and writer Ling Shuhua, is also associated with an intellectual circle, the New Moon Society. Her contradiction is that she believes in the Daoist "Art of Love", which to her intellectual peers is just a feudal old nonsense. The arrival of the Englishman gives her the opportunity to put this theory into practice.

And practice they do, quite a bit, and it's sensitively and sensuously described in the novel, even in the English translation I read, which is by Nicky Harman and the author's husband Henry Zhao. The eroticism is, of course, a problem for some people in China and in the UK, and so it came to pass that Ling Shuhua's daughter sued the author for libel in Chinese courts for defamation of the dead, and eventually succeeded in having the book banned in mainland China.

It hasn't quite been banned in the UK, but I'm getting the impression that it has been ignored on purpose. I find it quite shocking that I couldn't find a single review of the book. The English edition was published in 2002, so if it has been reviewed, the reviews should be on the web. Probably people perceived it like the subject's nephew, whose name is also Julian Bell, who didn't object to its publication but compared it to "black lace" type genre fiction.

Maybe it takes readers with intercultural sensitivity to appreciate this, but this is definitely not black lace material (and unlike Julian Bell, I have read black lace novels, I even know somebody who writes them). K really has something to tell us about what happens when cultures collide. The culture clash proves a bit too much for the English protagonist, who concludes towards the end of the book: "The fanatical love of this Chinese woman, like the violence of the Revolution, and everything else Chinese, was simply too alien for him to comprehend or accept."

I'm worried that this may be true for the British readers as well. I discovered the book in a charity shop, clearly unread. Somebody missed out on an amazing intercultural experience.

It's quite a tale, but I'm kind of detached from it

Daughter of the river
Hong Ying
Bloomsbury 1998

Prose and passion 19.1.2009

Having read K-the art of love recently, I became very curious about its author and decided to read her autobiography next. It didn't exactly tell me how she came to write K, nor how she came to write at all, but it turned out interesting nonetheless.

Hong Ying was born at Chongquing in 1962, at the end of a major famine that hit China as a result of population growth and mismanagement of the agricultural production. Thus, her novelized recollection of her childhood, culminating in discoveries about herself that she makes on her 18th birthday, could have easily produced one of those misery memoirs that have been so epidemic in recent years.

What saves it from descending that road is the curiously detached voice of the author who never seems to pity herself or the other protagonists. She describes hunger, violence and the regular sight of dead bodies floating down the river with equal emotional detachment, making the reader wonder whether this is a natural defence that children develop when they grow up in horrific circumstances, or whether there is an Asperger gene or two at play. For a teenage girl, our heroine cares remarkably little about what other people think. She sometimes even muses about her own detachment and aloofness.

Driven by her quest to solve the mysteries of her past, this story is quite gripping, even though the signposting often gives the events away beforehand. Most of all, it makes readers born at around the same time in a different place (like me, for example), consider ourselves very lucky indeed. What we now need from her is a sequel telling us how she escaped from the life of misery that could have become her destiny, and how she became a writer recognized around the world.

Elective affinities updated

Seduce me
Megan Clark
Kensington Books 2008

Prose and passion 26.1.2009

In his short novel "Wahlverwandtschaften" (Elective affinities), Goethe conducted an almost chemical experiment, exposing the established bond of Charlotte and Eduard to the affinities of new arrivals Otto and Ottilie. The four people rearrange to form two new couples, but it all remains quite platonic and then ends in tragedy.

In her 21st century update on the subject, US author and former exotic dancer Megan Clark explores the area where Goethe didn't dare to venture. She investigates the chemical rearrangement reaction (her version could even be described as a redox reaction, as a lot of fiery oxidation potential is transferred from one participant to another) in full carnal detail. As in chemistry, stability and reactivity are mutually exclusive but can be traded. One of her characters (but not the woman in the initially stable couple) is called Charlotte, too, but that was the only nod to the old master that I spotted.

The book makes good on the promise on the cover ("an erotic novel"), but it becomes clear that the eros serves to illustrate the issues of liberty and commitment and is neither the main issue nor the driving force. As in Goethe's novel, the structure and the momentum of the story is provided by the chemical rearrangement between the four characters.

Clark records this experiment with verve and a fast pace. What lets her down is the lack of words in the English language to describe erotic experiences, due to the long established taboos surrounding this area. Where Spanish has hundreds of words, English writers can use about 5. I think they should be more audacious in making up new ones. Go on, don't be shy! Also, there should be awards for erotic novels in English, like the "sonrisa vertical" award in Spain. So far, there is only a negative award for bad writing on this matter, which I believe is in itself very revealing.

Beware of orange trees

La mujer habitada
Gioconda Belli

Prose and passion 2.6.2009

I got an orange tree for my birthday last year, and some time after that, seemingly unrelated, I started reading La mujer habitada by Gioconda Belli. I've had the book for 10 years, but only got round to reading this, her debut novel, now, after her other major works. It soon turned out that the heroine is becoming "inhabited" by the spirit of a native American from the time of the Spanish conquest, and that the spirit had previously inhabited the orange tree she put in her patio. The spirit gradually turns her into a guerilla fighter. So I am now being very careful not to get too close to that tree of mine. Maybe it already tricked me into reading the book.

But seriously, and leaving any irrational fears of citrus plants aside, I liked the novel a lot, and it was particularly intriguing to read it soon after her more recent novel, El pergamino de la seducción (the scroll of seduction). Although the books are very different and set in different countries (Nicaragua and Spain, respectively), one could argue that they both feature as the main characters a young woman and a large house. In both books the young woman establishes a magical link to a woman who lived around 400 years earlier, via the orange tree in one case, and via the old house and its inhabitants in the other.

It's also interesting to read the book as a look back across time on the regime that the Sandinista movement swept away, knowing as we know now, that a few of the dashing young guerrilla leaders eventually became corrupted by power.

What I really don't understand is why her work is so underappreciated in the English-speaking world. Maybe it doesn't translate well?

Renaissance Man

The mercurial emperor – the magic circle of Rudolf II in renaissance Prague,
Peter Marshall

Prose and passion 24.7.2009

Imagine the most powerful political leader in the world decides he's not interested in politics and wars and all that and prefers to dedicate most of his time to collecting art and dabbling in science. Amazingly, it really happened: the leader in question was the Habsburg emperor Rudolf II, who presided over the Holy Roman Empire in the twilight years before central Europe drowned in the Thirty Years War.

In his very readable biography of the ruler who couldn't be bothered to rule, Peter Marshall follows the interests of his subject by dedicating only a little bit of space to the politics, and focusing on the art and the science instead. To get the politics out of the way first, many have slated Rudolf as a hopeless leader, but Marshall tends to support the view that his inaction and openness for divergent opinions (especially on religion, where he refused to support hardline Catholicism) quite possibly delayed the inescapable disaster by several decades.

Rudolf's credentials are much clearer in art and in science. His support for the most exciting astronomers of the day brought together the last and greatest naked-eye observer of the heavens, Tycho Brahe, and the best theoretician of the time, Johannes Kepler. Without Rudolf's patronage, the movement of the planets (Kepler's laws) might have remained unsolved for much longer.

Astronomers of the day still very much believed in astrology, or at least used it as a welcome source of income, so we're looking at an important junction between the medieval world views we now call superstition and the emerging modern science. Thus there are also alchemists and magi like the Briton John Dee populating the pages of this book.

In art, Rudolf supported many great artists, including Arcimboldo, who famously portrayed the emperor as a jigsaw made of vegetables. Marshall says that Rudolf often left important political figures waiting for an appointment, as he preferred to spend his time in the workshops of his artists, discussing their current work.

Of equal importance, though, was his unprecedented activity as a collector. He sent expert buyers including Jacopo Strada and his son Octavio to Italy, to buy art of what we now call the late Renaissance and bring it to Prague, the city which owes the time of its greatest glory to him.

With chapters dedicated to subject areas (magic, art, old astronomy, new astronomy) and the key figures representing them, Marshall's book is a very accessible read. Minor moans from my part include the fact that the Strada family have not been given a whole chapter as they would have deserved (I have a vested interest here, but I think it's fair to say that they are as important for Rudolf's circle as the visitor John Dee). I would have also wished the author to be more aware of points where the textbook history may be wrong: He uncritically refers to "Juana la loca" (Rudolf's great grand-mother in two lineages) as having been mad, and to a strained bladder as the cause of Tycho's death. In both cases, I would rather take side with the alternative views (see the review of El pergamino de la seducción, above).

The failure to acknowledge the possibility of Tycho's murder is especially unfortunate as Marshall reveals further links between Tycho and Shakespeare's Hamlet: The astronomer, whose research centre at Uraniborg was within eyesight of Hamlet's Elsinore, had two cousins named Rosenkrantz and Guildenstierne. Which rather fits in nicely with the murder case that Anderson makes and that Shakespeare may have known about. As the bard said, there are more things in heaven and on earth …

Seeing the funny side of failure

Verre Cassé
Alain Mabanckou

Prose and passion 10.9.2009

Although officially labelled a novel, Verre Cassé is essentially an amalgam of the life stories of the people washed up in the bar "Le Crédit a voyagé", somewhere in Congo (Brazzaville) as told by the most faithful customer, nick-named Verre Cassé (Broken Glass). Mimicking the unbridled speech flow of someone who had a few glasses and frustrations too many, Mabanckou writes without using a single full stop. Each chapter is a single rambling sentence, starting with a lower case and ending without any punctuation mark. This makes it a tad difficult to follow initially (as it would be difficult to follow the life story of a drunkard as told by himself down the pub past midnight), but I got used to it over time. I just had to adjust my attention span to the length of these uninterrupted rambles. (Note that I heroically resisted the temptation to write this review in the same style!)

While the book has its sad and depressing moments, on account of all the miscellaneous failures that led the characters to end up in this dump (and which they, to the last man, blame on the women in their lives), it can also be very funny a lot of the time. Early on, the scene where a committee of government officials searches for a catch-phrase for their boss, going through a whole dictionary of (unsuitable) citations, is quite hilarious, as are some of the events in and around the bar.

Citations occur not only in that scene, but throughout the book as a hallmark of our narrator. They come so thick and fast that I am wondering how on earth the translator of the English edition (which appeared a few months ago) got out of this problem. There are allusions to francophone literature en masse, both from France and from

French-speaking Africa, but also to French chanson, from Brassens to modern times. You don't have to have an education in French literature and culture to enjoy this book, but it would help enormously. Failing that, one can always read it for the drunken antics including a pissing contest.

Every picture tells a story

Luna cornata
Elvira Valgañón
Sobrelamesa Ediciones 2007

Prose and passion 17.9.2009

I discovered this charming little book in an Oxfam shop. It is about two people dreaming up stories inspired by old photos they get from an antiquarian bookshop in Dublin. These days, as the traditional, chemical process of making photos via negatives is disappearing rapidly, it is especially poignant to immerse oneself in the time when monochrome photos were at the cutting edge of technology, and having your family portraits taken at the local photographer's studio was an obligatory status symbol.

Making up life stories for the people portrayed in such photos opens a window into the social history of Dublin in the early 20th century, and, via migration, also to other parts of Europe. A certain distance is provided by the fact that the storytellers are foreigners in Dublin (she a photographer from Spain, he a visiting professor), looking at the Irish way of life with a degree of naïve admiration.

After reading this book, I noticed that there are in fact lots of old family portraits for sale on our local antiques market. Isn't it a shame that these have become dissociated from the stories that led to their creation and those that may have followed it? Making up stories is fun, and in this case makes for an inspiring read. Did the author really buy these photos in Dublin, or did she invent them, I wonder.

However, it would be even better if such photos could be re-attached to their real-life stories. For every old photo ending up in a car boot sale, there may be a genealogist who has the matching story and who would be overjoyed to find the photos to go with it.

But I'm digressing. Nice little book. Wonder what the author did next.

Exploring Bohemia

Among the Bohemians: Experiments in living 1900-1939
Virginia Nicholson
Penguin paperback 2003

Prose and passion 5.10.2009

Virginia Nicholson describes the citizens, customs, and traditions of a country that doesn't exist, but which is instantly recognisable: Bohemia. Using the lead metaphor of a virtual country of which Bohemians of the early 20th century were the inhabitants, she gives us a very detailed and colourful account of what life was like in that virtual place and real time.

Nicholson is, of course, supremely qualified to serve as Bohemia's ambassador to the modern world, as she descends from a family of Bohemian celebrities. She's not called Virginia for nothing: Virginia Woolf was her great-aunt, Woolf's biographer Quentin Bell her father, and Woolf's sister, the painter Vanessa Bell, her grandmother. (Oh, and Julian Bell, whose adventures in China inspired Hong Ying's book "K – the art of love", was Quentin's brother. What a family.)

Considering these family connections, it would have been easy for her to explore the faraway realm of Bohemia through the eyes of its best-known citizens. Instead, she turned her attention to the lives and times of the less successful artists and writers, to those for whom the move from "normal" England to Bohemia often meant a descent into real hardship.

Like an anthropologist describing the customs of a remote pacific island population, Nicholson studies the minute details of everyday life in Bohemia and devotes separate chapters to elementary aspects of the lives of the natives, including money, love life, education, interior decoration, clothes, food, housework, travel, and partying. Being an anarchistic place by definition there is no section on hierarchy or power.

These chapters are populated with a swirling cast of painters, models, and writers ranging from the almost famous to the long-forgotten, so the "Dramatis Personae" in the appendix becomes the most valuable part of the book, as one gradually learns to navigate the social networks of 1920s Bohemia. There have been many memoirs and biographies of the people involved, so there is no shortage of brilliant quotes to liven up the proceedings. My favourite one has to be the (unattributed) geometric description of the Bloomsbury set as a circle of friends who lived in squares and loved in triangles.

As additional material to make the sheer visual wealth of Bohemia more palpable, I recommend to consult the book "Charleston: a Bloomsbury house and garden" (Frances Lincoln 2004) as well, which Nicholson wrote together with her father, and which contains many wonderful photos by Alen Macweeney of the country house where Vanessa Bell and various associates lived. Additional visualisation help is provided by the movie Carrington (about the various love triangles involving the painter Dora Carrington and the writer Lytton Strachey), which is now available on DVD.

In all of this, the very endearing Bohemian philosophy of life shines through, which essentially implies that for each individual their art, love, and friendships are infinitely more important than the ten million rules and regulations that Victorian society imposed on its members. Living in a society which in many ways is closer to Bohemia's jurisdiction than to Victoria's, we are bound to find the Bohemian rule-breaking and mischief-making amusing, even in instances which contemporaries must have considered truly shocking.

Therefore, the whole makes for a highly entertaining read, but it also provides food for thought and self-questioning. Many of Bohemia's battles against Victorian norms have been won by now – for instance, women can wear their hair short, and men theirs long without shocking anybody. And yet, people with any kind of creative inclination still face the age-old dilemma whether they can afford to put their art (in the widest sense) first, or whether they have to make bourgeois compromises to ensure they can pay their bills. Today, there isn't a separate country called Bohemia any more. On the sliding scale between Bohemian and Bourgeois, everybody has to find their own place.

Public enemy

Ulrike Meinhof: Die Biographie
Jutta Ditfurth
Ullstein 2007

Prose and passion 7.10.2009

Ulrike Marie Meinhof, who would have turned 75 today, was West Germany's public enemy No. 1 from a fateful day in May 1970 through to her death in a purpose-built high-security prison six years later, and indeed well beyond the grave. Together with Andreas Baader and Gudrun Ensslin, she was seen as the co-founder of the organisation that was to develop into the Rote Armee Fraktion (RAF), built on the mistaken belief that it could trigger a revolution by starting urban guerrilla warfare against a state they saw as increasingly oppressive. The group killed a total of 34 people, and its own casualties, combined with the random ones killed by police forces going over the top, were on a similar scale. What it eventually achieved was exactly the opposite of what it set out to do – it made the state more oppressive. Many democratic rights and civil liberties were cut and undermined by the social democrat / liberal democrat governments of the 70s striving to be seen as tough in their fight against terrorism.

With all the panic and witch-hunting that marked these "years of terror", few people realised that the biographies of the failed revolutionaries were completely at odds with the public image spread by government and media. The RAF was probably the only political organisation of its time that was dominated and led by women. Among the male recruits, there were adventurers who got a kick out of the idea of playing war with real guns (the name Andreas Baader springs to mind). Most of the members, however, and the women especially, came to the movement from social engagements, either working with marginal groups of society, or, in the later stages, motivated by the ill treatment of the first generation of RAF prisoners.

Ulrike Meinhof also had a track record of helping the weakest and most marginalised groups of society (e.g. teenagers in care homes), but beyond that she also was one of the country's most talented political journalists. As a columnist for left-wing magazine Konkret, she was the most eloquent critic of German politics in the 1960s, from the remilitarisation through to the "Notstandsgesetze".

How somebody of that intellectual standing, who commands a significant audience, can turn their back on society and take up arms in a struggle that was doomed from the start is a very important question that justifies a major biography. While there have been shorter volumes on Meinhof (e.g. by Peter Brückner, Mario Krebs), Jutta Ditfurth's magnum opus is much more ambitious in scope.

Herself a lifelong dissident, the daughter of science writer Hoimar von Ditfurth dropped the aristocratic "von" from her name and became a founding member of Germany's Green Party, which she left after a few years as it drifted away from the radical idealism of its founders. With hindsight, it makes perfect sense that Ditfurth should have been drawn to this subject.

And what a success she made of it. It shows on every page that, while she doesn't agree with the violent path that Meinhof took, Ditfurth does understand perfectly well what made her tick, and how she felt that society had taken a wrong turn. She uses the pivotal liberation of Andreas Baader in 1970, after which Meinhof had to live undercover, as an opening scene, she tells the life essentially in chronological order, from the background of her parents, who both died young and left Ulrike in the care of her mother's friend, the academic Renate Riemeck, through to the prison years and the controversial death, which the authorities hastily labelled a suicide. Conspiracy theories implying that she was murdered in her cell at Stuttgart-Stammheim have never quite been quelled, and Ditfurth returns an open verdict on this prickly question.

The family and political networks that shaped her life before 1970 come alive on these pages and make make the ensuing tragedy all the more dramatic. There are gaps in the post-1970 knowledge, which are inevitable because of the clandestine nature of the RAF operations, which have never been unravelled in court. (Typically, all members caught were convicted on the basis of being members of a terrorist organisation by the very same judges that had previously rejected any guilt to be associated with membership of Nazi organisations, one of the many bitter ironies in this story.)

Jutta Ditfurth has delivered the definitive account of the life of somebody who could have been post-war Germany's most admired intellectuals and instead turned into its most-hated (which, of course, requires a great deal of courage from the author, as some of that residual hatred will surely be reflected in her direction). The poet Erich Fried called Meinhof the most important German woman since Rosa Luxemburg. The way both women perished should concern us all.

Intimate adventures of a scientist

The intimate adventures of a London call girl
Belle de Jour
Weidenfeld & Nicolson 2005

Prose and passion 27.11.2009

This book is a bit of a cause célèbre in the UK but readers elsewhere in the world may not know that it's the intermediate between a very successful early noughties blog and a somewhat controversial TV series starring Billie Piper, a former teen pop star and Dr Who assistant. I found a copy at Oxfam a while ago and read the first 50 pages or so, but then something else must have been more urgent and I forgot all about it. I picked it up again after the revelation that the author is a working research scientist, Dr Brooke Magnanti, with a PhD in cancer epidemiology.

Intrigued by the strange case of Dr Magnanti and Ms. Jour (presumably she must have had a third name for the agency work?), I finished the book off quite quickly, and found it was even more fascinating to read as "The intimate adventures of a London scientist" rather than as those of an anonymous call girl.

With hindsight, it really is quite obviously the work of a scientist. In a section about comforting an acquaintance over a breakup, she concludes: "… I felt for her. I've been on both sides of that equation." Similar science-inspired expressions and observations are pop up repeatedly. She often refers to her university years, though she doesn't quite tell us what she studied and where. Or does she?

In fact, one section that is quite hilarious to read post facto is a list of "Pub Games for Whores." One such game, designed to confuse men who try to chat you up, is to invent an "implausible occupation." The paragraph ends:

Extra points if he actually holds that job. 'Really? You're an epidemiologist? What a coincidence!' (p187)

Coincidence, indeed. The truth is in there. Also, the tabloids needn't have bothered to chase her poor old dad. It's in the book:

Have I mentioned that my father is an embarrassing perv? Runs in the bloodline, I suppose. (p168)

Reading the book as the work of a fellow scientist makes it quite endearing. We have all shared the same career worries at some point in our lives, after all. Although there are very few scientists who can write as well as she does. Clearly, this woman has many talents – research experience, communicates well, can deal with people even in awkward situations – so, from a society point of view, I find it troubling that she was underappreciated and unemployed for long enough to even consider prostitution.

I was glad to hear that, having explored other career paths, she's sticking with science although her publishing success alone would probably pay the bills quite nicely. "Working in science is important to me" Magnanti told New Scientist in an interview after coming out. So good luck to her, I'm sure she will do well, and I hope that one day she'll write about science too.

Orient meets Occident

The enchantress of Florence
Salman Rushdie
Jonathan Cape 2008

Prose and passion 21.12.2009

First up, I have to admit that this, Rushdie's tenth novel, is the first one I've read, so I may end up saying things that are blatantly obvious to people more familiar with his work. I'm also quite happily ignorant of the work of most of his male, British-based contemporaries. So I'm in the fortunate position of reading the book purely on the merits of its contents.

What attracted me to the novel was the fact that it makes surprising connections between Orient and Occident, weaving history and magic realism into a very colourful pattern, which, I think, tells us something about our shared cultural heritage. The enchantress of the title is Qara Köz, (supposedly) an aunt of the Mughal emperor Akbar the great. While reading the book, I wasn't sure whether she was among the historic or the fictional characters, but just now a google search for her name brought up only references to the novel, so I guess she's an imaginary person (just like Akbar's favourite wife is imaginary within the novel).

At the court of Akbar arrives a European traveller who claims to be her son, although he is obviously too young for this to be true. The stranger tells an elaborate story to back up his claim, leading Qara Köz from the Mughal empire to Persia, then to

Florence, where she meets historical figures including Niccolo Machiavelli, Andrea Doria, and the Medici princes, then to the New World, where the story-teller is born, who completes her circumnavigation of the globe.

Orient and Occident in Rushdie's historical fantasy are equally violent and Machiavellian places, where nobody can trust anybody else. Celebrated military leaders typically die from treason within their own ranks. Everybody has to watch their back constantly, and the "enchantress" has to switch her allegiance repeatedly and walk the very fine line between being enchanting and being persecuted as a witch.

It is fascinating to see a synopsis of the histories that are normally treated as though they had happened on different planets, even though there definitely were trade connections, and thus must have been people who moved between the different cultures flowering in distant parts of the world.

I generally think that Europeans tend to be in denial about the influence that oriental culture, promoted by the Arabs in Spain and by the Turks at the Eastern borders of Christian Europe, had on our cultural evolution. Too keen to define itself as the Christian Occident against the Islamic Orient, Europe failed to appreciate the good things the Islamic world had to offer, e.g. Islamic science in the Dark Ages, when there was not much European science happening at all.

Therefore, this tale of two cultures projects a very interesting light on renaissance history (familiar in the European part, and new to me in the Indian part). Of course one could also read it as a fable reflecting modern cultural conflicts.

www.ingramcontent.com/pod-product-compliance
Ingram Content Group UK Ltd.
Pitfield, Milton Keynes, MK11 3LW, UK
UKHW041830290726
14061UKWH00004BA/177/J